Automotive Engineering
Lightweight, Functional, and Novel Materials

Series in Materials Science and Engineering

Series Editors: **Alwyn Eades,** Lehigh University, Bethlehem, Pa., USA
Evan Ma, Johns Hopkins University, Baltimore, Md, USA

Other books in the series:

Strained-Si Heterostructure Field Effect Devices
C K Maiti, S Chattopadhyay, L K Bera

Spintronic Materials and Technology
Y B Xu, S M Thompson (Eds)

Fundamentals of Fibre Reinforced Composite Materials
A R Bunsell, J Renard

Novel Nanocrystalline Alloys and Magnetic Nanomaterials
B Cantor (Ed)

3-D Nanoelectronic Computer Architecture and Implementation
D Crawley, K Nikolic, M Forshaw (Eds)

Computer Modelling of Heat and Fluid Flow in Materials Processing
C P Hong

High-K Gate Dielectrics
M Houssa (Ed)

Metal and Ceramic Matrix Composites
B Cantor, F P E Dunne, I C Stone (Eds)

High Pressure Surface Science and Engineering
Y Gogotsi, V Domnich (Eds)

Physical Methods for Materials Characterisation, Second Edition
P E J Flewitt, R K Wild

Topics in the Theory of Solid Materials
J M Vail

Solidification and Casting
B Cantor, K O'Reilly (Eds)

Fundamentals of Ceramics
M W Barsoum

Aerospace Materials
B Cantor, H Assender, P Grant (Eds)

Series in Materials Science and Engineering

Automotive Engineering
Lightweight, Functional, and Novel Materials

Edited by

Brian Cantor
University of York, UK

Patrick Grant
Oxford University, UK

Colin Johnston
Oxford University, UK

Taylor & Francis
Taylor & Francis Group
New York   London

Taylor & Francis is an imprint of the
Taylor & Francis Group, an **informa** business

FIRST INDIAN REPRINT, 2010

CRC Press
Taylor & Francis Group
6000 Broken Sound Parkway NW, Suite 300
Boca Raton, FL 33487-2742

International Standard Book Number-13: 978-0-7503-1001-7

Library of Congress Cataloging-in-Publication Data

Cantor, Brian.
   Automotive engineering : lightweight, functional, and novel materials / Brian Cantor, Patrick Grant, Colin Johnston.
     p. cm.
   Includes bibliographical references and index.
   ISBN 978-0-7503-1001-7 (alk. paper)
   1. Motor vehicles--Materials. I. Cantor, Brian. II. Grant, Patrick. III. Johnston, Colin. IV. Title.

TL154.C36 2007
629.2'32--dc22   2007015715

**Visit the Taylor & Francis Web site at
http://www.taylorandfrancis.com**

**and the CRC Press Web site at
http://www.crcpress.com**

Contents

Section 1   Industrial Perspective

Section 2   Functional Materials

Section 3   Light Metals

Section 4   Processing and Manufacturing

Preface

This book is a text on automotive materials, arising from presentations given at the fifth Oxford–York–Kobe Materials Seminar, held at the Kobe Institute on 10–13 September 2002.

The Kobe Institute is an independent non-profit-making organization. It was established by donations from Kobe City, Hyogo Prefecture, and more than 100 companies all over Japan. It is based in Kobe City, Japan, and is operated in collaboration with St. Catherine's College, Oxford University, United Kingdom. The chairman of the Kobe Institute Committee in the United Kingdom is Roger Ainsworth, master of St. Catherine's College; the director of the Kobe Institute Board is Dr. Yasutomi Nishizuka; the academic director is Dr. Helen Mardon, Oxford University; and the bursar is Dr. Kaizaburo Saito. The Kobe Institute was established with the objectives of promoting the pursuit of education and research that furthers mutual understanding between Japan and other nations, and to contribute to collaborations and exchanges between academics and industrial partners.

The Oxford–York–Kobe seminars are research workshops that aim to promote international academic exchanges between the United Kingdom/ Europe and Japan. A key feature of the seminars is to provide a world-class forum focused on strengthening connections between academics and industry in both Japan and the United Kingdom/Europe, and fostering collaborative research on timely problems of mutual interest.

The fifth Oxford–York–Kobe Materials Seminar was on automotive materials, concentrating on developments in science and technology over the next ten years. The cochairs of the seminar were Dr. Hisashi Hayashi of Riken, Dr. Takashi Inaba of Kobe Steel, Dr. Kimihiro Shibata of Nissan, Professor Takayuki Takasugi of Osaka Prefecture University, Dr. Hiroshi Yamagata of Yamaha, Professor Brian Cantor of York University, Dr. Patrick Grant and Dr. Colin Johnston of Oxford University, and Dr. Kaizaburo Saito of the Kobe Institute. The seminar coordinator was Pippa Gordon of Oxford University. The seminar was sponsored by the Kobe Institute, St. Catherine's College, the Oxford Centre for Advanced Materials and Composites, the UK Department of Trade and Industry, and Faraday Advance. Following the seminar, all of the speakers prepared extended manuscripts in order to compile a text suitable for graduates and for researchers entering the field. The contributions are compiled into four sections: industrial perspective, functional materials, light metals, and processing and manufacturing.

The first four and seventh Oxford–York–Kobe Materials Seminars focused on aerospace materials in September 1998, solidification and casting in September 1999, metal and ceramic composites in September 2000, nano-materials in September 2001, and spintronic materials in September 2004. The corresponding texts have already been published in the IOPP Series in Materials Science and Engineering and are being reprinted by Taylor & Francis. The sixth Oxford–York–Kobe Materials Seminar was on magnetic materials in September 2003 and the eight Oxford–York–Kobe Materials Seminar will be on liquid crystals in April 2008.

Acknowledgments

The editors would like to thank the Oxford–Kobe Institute Committee, St. Catherine's College, Oxford University, and York University for agreeing to support the Oxford–York–Kobe Materials Seminar on Automotive Materials; Sir Peter Williams, Dr. Hisashi Hayashi, Dr. Takashi Inaba, Dr. Kimihiro Shibata, Professor Takayuki Takasugi, Dr. Hiroshi Yamagata, Dr. Helen Mardon, and Dr. Kaizaburo Saito for help in organizing the seminar; and Pippa Gordon and Sarah French for help with preparing the manuscripts.

Individual authors would like to make additional acknowledgments as follows:

Chapter 3: We are grateful for the support of the UK funding agencies, the University of Reading, and Toyota for financial support for this work.

Chapter 4: We are grateful for the support of the UK funding agencies, the University of Reading, and Toyota for financial support for this work.

Chapter 7: The author wishes to acknowledge support from HITEN and the CEC Thematic Network Programme, and contributions from Riccardo Groppo, Fiat Research, Italy; Wolfgang Wondrak, Daimler Chrysler, Germany; and Wayne Johnson of Auburn University, United States.

Chapter 14: Support for this work has been provided by the Cambridge–MIT Institute (CMI). Andrew Cockburn of Cambridge University made some of the stiffness measurements and produced the 3-D array sheet. Sheets with flocked and mesh cores were provided by Jerry Karlsson of HSSA Ltd. Thanks are also due to Steve Westgate of TWI for extensive help with welding activities and to Peter Rooney and Lee Marston of FibreTech for ongoing collaboration related to supply of fibers and development of the processing technology.

Chapter 16: The author expresses sincere thanks to Dr. T. Tetsui at Mitsubishi Heavy Industries for the supply of some of the TiAl-based intermetallic materials.

Editors

Brian Cantor was educated at Manchester Grammar School and Christ's College, Cambridge. He has worked at Sussex, Oxford, and York Universities, and with leading companies, such as Alcan, Elsevier, General Electric, and Rolls-Royce. He is on the boards of White Rose, Worldwide Universities Network, Yorkshire Science, and the National Science Learning Centre; and was on the boards of Amaetham, York Science Park, Isis Innovation, and the Kobe Institute. He has advised agencies such as EPSRC, NASA, the EU, and the Dutch, Spanish, and German governments. At Oxford he was Cookson Professor of Materials, the first head of the Division of Mathematical and Physical Sciences, and a member of the General Board and Council. He was appointed in 2002 as vice-chancellor of the University of York.

His research investigates the manufacture of materials and has contributed to improvements in products such as electrical transformers, pistons, car brakes, aeroengines, and lithographic sheeting. He has supervised over 130 research students and post doctoral fellows, published over 300 papers, books, and patents, and given over 100 invited talks in more than 15 countries.

He was awarded the Rosenhain and Platinum Medals of the Institute of Materials, the first for "outstanding academic/industrial collaboration" and the second for "lifetime contributions to materials science." He is an honorary professor at Northeastern University Shenyang, Zhejiang University, and the Chinese Institute of Materials, and is a member of the Academia Europea, and the World Technology Forum and is on the ISI list of Most Cited Scientists. He is a fellow of the Institute of Materials, the Institute of Physics, and the Royal Academy of Engineering, elected to the Royal Academy as "a world authority on materials manufacturing."

Patrick Grant received a B.Eng. in metallurgy and materials science from Nottingham University in 1987, and a D.Phil. in materials from Oxford University in 1991. He was a Royal Society University research fellow and Reader in the Department of Materials, Oxford University, and became Cookson Professor of Materials at Oxford University in 2004. His published work of over 100 papers concerns advanced materials and processes for industrial structural and functional applications, especially in the aerospace and automotive sectors. He has been granted three patents licensed to industry.

He was director of the Oxford Centre for Advanced Materials and Composites (1999–2004) that coordinates industrially related materials at Oxford University and is currently director of Faraday Advance, a component of the Materials Knowledge Transfer Network, a government and industry funded national partnership that links the science base with industry in the field of advanced materials. Faraday Advance focuses on new materials—lightweight and low environmental impact materials for transport applications. He is a member of the 2008 Research Assessment Exercise Panel for Materials and a member of the Defense and Aerospace National Advisory Committee for Materials and Structures.

Colin Johnston splits his time as a technology translator with Faraday Advance—the Transport Node of the Materials Knowledge Transfer Network—and as coordinator of the Institute of Industrial Materials and Manufacturing section of the Department of Materials, Oxford University, where he has held the position of senior research fellow since 2001. He received a B.Sc. (Honors) in chemistry from the University of Dundee in 1984, followed by a Ph.D. in surface science and catalysis in 1987, also from the University of Dundee. In 1987 he joined AEA Technology at the Harwell Laboratory where he was a member of the Materials Development Division specializing in materials characterization. He later developed electronic materials for harsh environments, working on wide band gap semiconductors and microsystems. Johnston was operations manager of the Electronic Materials and Thermal Management business of AEA Technology from 1998 to 2000, when he assumed a post within the central corporate structure, managing innovation and new technology acquisitions for the company.

He is director of HITEN—the EU-funded network for high temperature electronics, where he established a pan-European strategy. He is also cochair of the U.S. High Temperature Electronics Biennial Conference Series and has published over 80 papers in scientific journals and edited several books on high-temperature electronics.

Contributors

Ed Allnutt
Crompton Technology Group Ltd.
Banbury, Oxon
United Kingdom

Will Battrick
Crompton Technology Group Ltd.
Banbury, Oxon
United Kingdom

Michael Bowker
School of Chemistry
Cardiff University
Cardiff, United Kingdom

S. W. Charles
Department of Physics
University of York
Heslington, York, United Kingdom

T. W. Clyne
Engineering Department
University of Cambridge
Cambridge, United Kingdom

Roger Davidson
Crompton Technology Group Ltd.
Banbury, Oxon
United Kingdom

Clifford M. Friend
Cranfield University
Cranfield, Bedfordshire
United Kingdom

Kenji Higashi
Osaka Municipal Technical
  Research Institute
Osaka Prefecture University
Nakaku, Sakai
Osaka, Japan

Takashi Inaba
Kobe Steel
Chuo-ku, Kobe
Hyogo, Japan

Hirofumi Inoue
Department of Materials Science
Osaka Prefecture University
Nakaku, Sakai
Osaka, Japan

M. Itoh
Shinko Wire Company Ltd.
Izumisano, Japan

Colin Johnston
Department of Materials
Oxford University
Oxford, United Kingdom

Mark Jolly
Process Modelling Group
University of Birmingham
Birmingham, United Kingdom

J. M. Kell
TWI Ltd.
Great Abington, Cambridge
United Kingdom

Mamoru Mabuchi
National Industrial Research
  Institute of Nagoya
Nagoya, Japan

A. E. Markaki
Engineering Department
University of Cambridge
Cambridge, United Kingdom

T. Miyoshi
Shinko Wire Company Ltd.
Izumisano, Japan

Toshiji Mukai
Osaka Municipal Technical
  Research Institute
Osaka Prefecture University
Nakaku, Sakai
Osaka, Japan

T. Mukai
Shinko Wire Company Ltd.
Izumisano, Japan

S. Nakano
Shinko Wire Company Ltd.
Izumisano, Japan

Kevin O'Grady
Department of Physics
University of York
Heslington, York
United Kingdom

Ivana K. Partridge
Cranfield University
Cranfield, Bedfordshire
United Kingdom

V. Patel
Department of Physics
University of York
Heslington, York
United Kingdom

Nik Petrinic
Department of Engineering Science
Oxford University
Oxford, United Kingdom

Kimihiro Shibata
Department of Materials Science
  and Engineering
Miyagi National College
  of Technology
Natori, Miyagi
Japan

Takayuki Takasugi
Department of Metallurgy
  and Materials Science
Osaka Prefectural University
Sakai, Osaka
Japan

Kazuhisa Toh
Mazda Motor Corporation
Kanagawa-ku, Yokohama
Kanagawa, Japan

Osamu Umezawa
Yokohoma National University
Division of Mechanical Engineering
  and Materials Science
Hodogaya, Yokohama
Japan

J. G. Wylde
TWI Ltd.
Great Abington, Cambridge
United Kingdom

Section 1

Industrial Perspective

Cars and automobiles are developing rapidly, with increasing global competition between industrial manufacturing companies, and with increasing social requirements for reduced noise and pollution, increased safety and energy efficiency, higher performance, and, at the same time, reduced cost. New materials and processing techniques are needed to underpin these developments. The industrial scene, the key design drivers, and the emerging new materials and processing technologies are covered in detail in this section.

Chapter 1 discusses the development of future vehicles and the associated new materials for a wide range of automotive components, concentrating on the importance of improved safety, reduced environmental damage, the role of information processing, and the overarching need for cost-effectiveness in a competitive market. Chapters 2 and 3 concentrate on more specific issues. Chapter 2 describes the development of suitable aluminum alloys and associated processing techniques to manufacture lighter body panels, with improvements in energy efficiency, fuel savings, and performance. Chapter 3 describes the development of a variety of different polymer composites and their associated moulding techniques to make stronger and more effective module carriers, which are used to allow rapid and cost-efficient manufacture of complex multiple parts.

1

Future Vehicles and Materials Technologies

Kimihiro Shibata

CONTENTS

Introduction

In the twenty-first century, cars should be designed and engineered to be in harmony with people and nature. Environmental and safety issues today call for technological improvements. Reduction of CO_2 emissions and improvement of fuel economy can be achieved together with crashworthiness through contributions made by material technologies. Besides improving mechanical properties and cost competitiveness, peripheral technical issues, such as forming and joining technologies, and environmental performance, should be addressed prior to the deployment of a new material. Cooperation among material suppliers, parts suppliers and carmakers, or among carmakers themselves, in a simultaneous or concurrent manner, is becoming more important than ever.

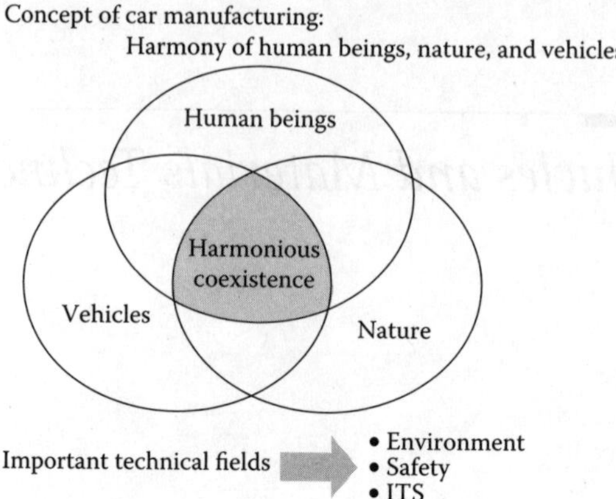

FIGURE 1.1
Concept of harmonization.

More than a century has passed since the automobile was invented, and the environment surrounding the automotive industry has undergone a lot of changes on countless occasions in the intervening years. Notable changes started with the introduction of mass production technology that was established for the Ford Model T series in the 1910s. After World War II, Japanese carmakers resumed passenger vehicle production and began to pursue quality improvements. The two oil crises in the 1970s promoted the development of low fuel consumption technologies. Following the two oil crises, stricter exhaust emission regulations were enforced and intense competition to secure higher levels of performance unfolded in the early 1990s. Since the latter half of the 1990s, the focus has been on safety and environmental issues. In line with this progression, the concept of harmonious coexistence, which is striking a balance among human beings, nature and vehicles, is expected to increase in importance in vehicle manufacturing in the twenty-first century. Important technology fields for achieving this harmonization are the environment, safety, and intelligent transportation systems (ITS), as indicated schematically in Figure 1.1.

This chapter surveys the social conditions surrounding the automotive industry. An overview of the history of automotive materials will then be given, followed by a discussion of projected future trends in material technologies.

Environmental Issues

Protection of the global environment, which includes conservation of resources, is a pressing issue. Figure 1.2 shows the increase over the last 50 years in the global number of vehicles.[1] In 1950, 70 million vehicles were on the road in relation to a world population of 2.4 billion people. By 2000, the number of

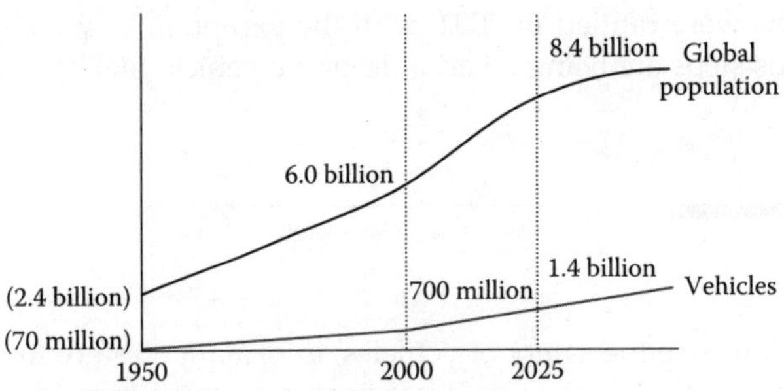

FIGURE 1.2
Number of vehicles and global population.

vehicles had increased to 700 million, while the world population had grown to 6 billion. In other words, the number of vehicles increased tenfold over the last 50 years of the twentieth century: It is estimated to double to 1.4 billion by 2025. With this increase in the number of vehicles, oil consumption has continued to rise, and environmental issues have become more serious.

The possibility has been pointed out that global oil production might peak in the year 2015 and begin to decline after that.[2] Therefore, there are strong demands for the conservation of oil resources. Countries around the world have adopted standards that regulate the allowable levels of hydrocarbons (HC), carbon monoxide (CO), and nitrogen oxides (NO_x) in vehicle exhaust gas. These exhaust emission regulations will be further tightened in the future. Furthermore, carbon dioxide (CO_2) in exhaust emissions has been singled out as one of the causes of global warming. The Kyoto Protocol set targets for reducing CO_2 emissions. To achieve the targets set for Europe, the United States, and Japan in 2010, the CO_2 emission level of cars with a gasoline engine needs to be reduced by 6%–8% compared with 1995 models. This means that their average fuel economy must be improved by 25%,[3] as shown in Table 1.1.

TABLE 1.1

COP3 Targets for Reducing CO_2 Emissions and Improving Fuel Economy[3]

	CO_2 Reduction (vs. 1990)[*1]	Fuel Economy
Japan	6%	Passenger car with gasoline engine: improved by 23% (by 2010 vs. 1995) ±15 km/L Passenger car with diesel engine: improved by 15% (by 2005 vs. 1995) ±12 km/L
Europe	8%	Passenger vehicle: improved by 25% (by 2008 vs. 1995) CO_2:140 g/km
USA	7%	Passenger car CAFE target: 27.5 mpg (after 1990) (PNGP project is under way.)

[*1] Period: 2008–2012.

These targets were ratified in 2002, with the exception of the United States, and vigorous steps are being taken to improve vehicle fuel economy.

Safety

In order to improve the safety of vehicles, information safety for preventing accidents in addition to crash safety is becoming more important, as shown in Figure 1.3. In the course of developing technologies for improving crash safety, traffic accidents are reproduced and analyzed. The results of these analyses have been applied to develop new crash safety technologies, such as an automatic braking system for reducing the collision speed, and an emergency stopping system. In the area of information safety, advanced safety vehicles and advanced highway systems are being developed using sophisticated technologies like intelligent vision-sensing and car-to-car communication systems.

In recent years, the results of car crash tests conducted under a new car assessment program (NCAP) in various countries, as well as the accident rates of individual car models, have been disclosed. Such data are usually considered in the determination of car insurance premiums. Due to stricter safety regulations and the disclosure of information regarding safety, consumers are more concerned about safety today than ever before. Based on analyses of traffic accidents, the new car assessment program will continue to adopt more precise and sophisticated collision tests. Various new car assessment tests and regulations concerning crash safety are being prepared for implementation in the coming years, as shown in Figure 1.4.

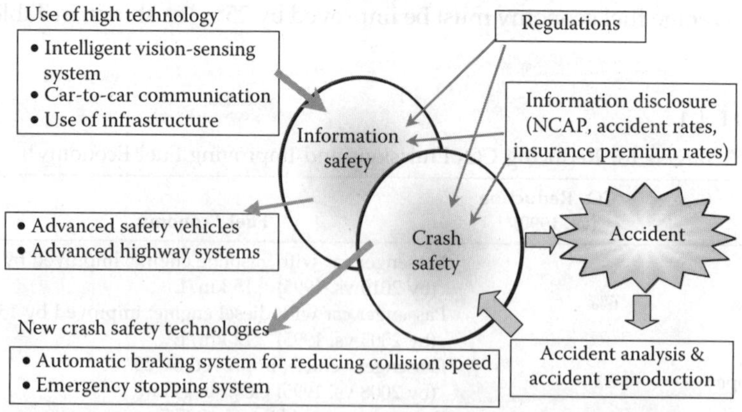

FIGURE 1.3

Vehicle crash safety and information safety.

(NCAP : New Car Assessment Program)

		2000		2005	
NCAP	Japan	Full overlap frontal Side impact Brake performance	Offset frontal Overall evaluation CRS evaluation	Pedestrian protection Head rest (dynamic)	
	USA	Full overlap frontal Side impact (Offset frontal (IIHS))		Roll-over avoidance Whiplash evaluation (dynamic) Brake performance	Offset frontal Enforced side impact (compatibility)
	EU	Full overlap frontal Side impact Pedestrian protection CRS evaluation		Brake performance	Full overlap frontal Frontal (compatibility) Whiplash evaluation (dynamic)
Safety regulations				Advanced airbag (USA)	Pedestrian protection (J, EU) Advanced headlamps (J, US, EU)
				International standardization ⟵⟶	

FIGURE 1.4

Trends in NCAP tests and safety regulations in Japan, the United States, and the European Union.

Intelligent Transportation Systems (ITS)

Intelligent transportation systems (ITS) are highway traffic systems in which smart vehicles and smart roads are integrated. These systems are expected to improve transport efficiency and safety, make driving more enjoyable, and also contribute to environmental protection, as shown in Figure 1.5. For example, CO_2 and NO_x levels would be markedly reduced if the average driving

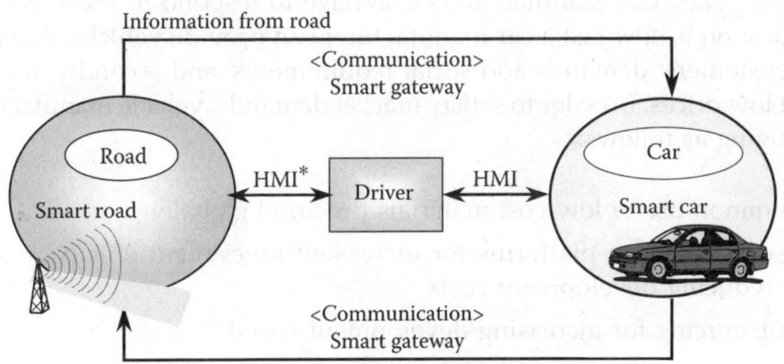

FIGURE 1.5

Intelligent transport systems.

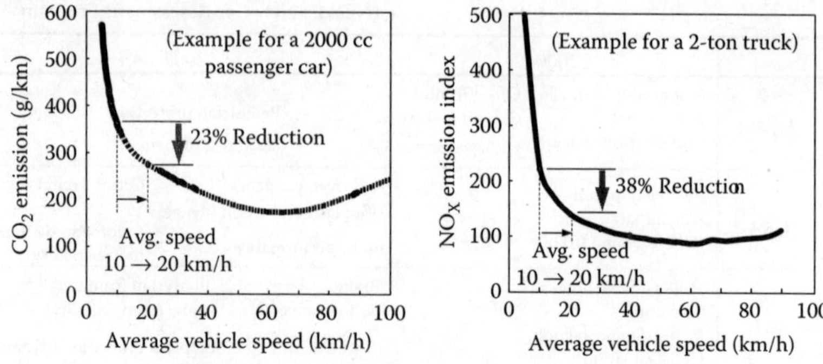

FIGURE 1.6
Emission levels as a function of average vehicle speed.[4]

speed during traffic congestion could be increased from 10 to 20 km/h through the use of an intelligent transportation system,[4] as shown in Figure 1.6. Moreover, the number of traffic accidents might also be reduced, for example, by applying an adaptive cruise control system together with intelligent transportation system capabilities.

Market Trends

Customer needs are becoming greatly diversified, and the speed at which they are changing is accelerating. During Japan's bubble economy in the late 1980s, customers preferred luxurious products of a uniform style, but vehicles having good cost performance and individuality have been well received in recent years. Car manufacturers also have to respond to social issues. A key question is how fast a car manufacturer can provide vehicles that firstly meet customers' demands and social requirements, and secondly are available at low prices. In order to satisfy market demands, vehicle manufacturing is changing as follows:

Common use of low-cost materials procured globally

Use of common platforms for increasing investment efficiency and reducing development costs

Outsourcing for increasing development speed

These changes in vehicle manufacturing are undermining the traditional "keiretsu" system of company groupings in Japan. Today, automobile parts are assembled into modules by suppliers and provided to car manufacturers,

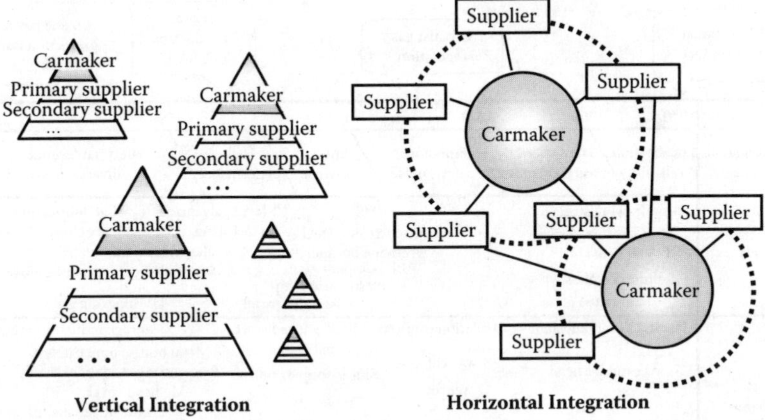

FIGURE 1.7
Alternative types of company grouping.

and it is not unusual nowadays for rival carmakers to purchase parts from the same parts supplier. The traditional vertical integration of companies is changing to more horizontal integration, as indicated in Figure 1.7. This horizontal integration is basically composed of "give & take" relationships. The idea that everything should be done in-house or by "keiretsu" companies has vanished. In this new structure, global networks for information, cooperation, and human resources are becoming very important elements of corporate competitiveness.

Automotive Materials

Figure 1.8 shows a history of automotive, mainly metal, materials. Over the years, new materials have been developed along with changes in social conditions and market requirements.

Car Body Materials

New materials for the car body have been developed to improve corrosion resistance and to reduce vehicle weight. In the 1950s and 1960s, mass production technologies were developed because of higher vehicle demand. High performance and reliability were also the market trends at that time. Deep drawing steel sheets with good formability were developed in the 1950s, followed by the development of anti-corrosive steel sheets in the 1960s. In the 1970s and 1980s, low fuel consumption was a keen issue because of the

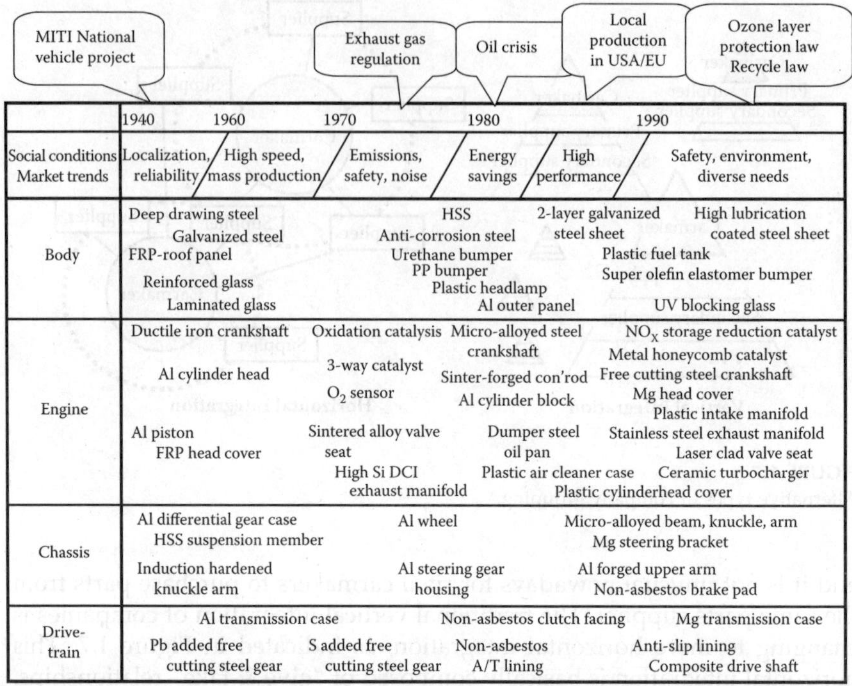

FIGURE 1.8
New materials used in vehicles.

two oil crises. High-strength steel sheets were developed in response to this issue and have contributed to lightening vehicles by reducing sheet thickness. In the 1990s, safety and environmental issues became primary concerns in the automotive industry, and further work was done on developing technologies for weight reductions. Aluminum alloy sheets were developed in this connection and applied to various body panels such as the engine hood, and have contributed to achieving lighter vehicles.

Materials for Engine Components

New materials for engines have been developed to improve engine durability and performance as well as to reduce the weight of components. In the 1950s, ductile cast iron suitable for volume production was developed and applied to crankshafts. In the 1980s, micro-alloyed steels were developed and applied to crankshafts and connecting rods. Sinter-forged connecting rods were also developed. For the sake of weight reductions, aluminum alloys were used for cylinder heads, and stainless steels for exhaust manifolds. In the 1990s, aluminum alloys were applied to cylinder blocks, and magnesium alloys to cylinder head covers.

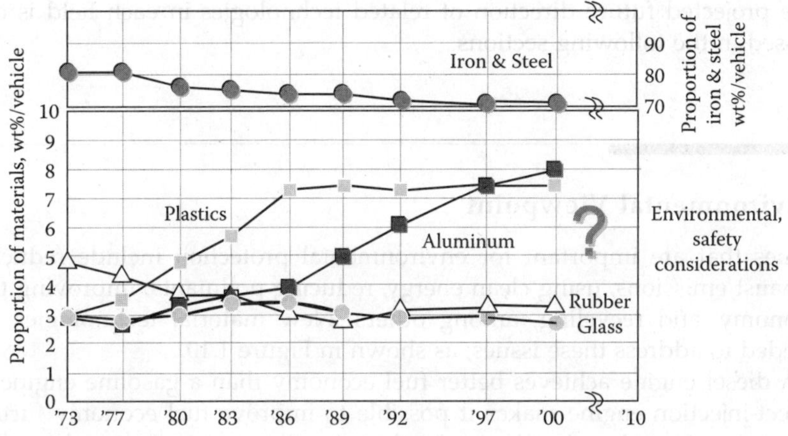

FIGURE 1.9

Material composition of a typical passenger vehicle.

Materials for Chassis and Powertrain Components

New materials for chassis and powertrain components have been developed mainly to improve durability and reduce weight. High-strength steel sheets were applied for suspension members and aluminum alloys for wheels. Knuckles, arms, and I-beams made of micro-alloyed steels were developed. Aluminum alloys are now being used for transmission cases. Gears are made of free-cutting steels. In recent years, magnesium alloys have been applied to steering system components and transmission cases. Carbon composites with fiber-reinforcement have begun to be used for propeller shafts.

A breakdown[3] of the materials used in a typical passenger vehicle for the Japanese market is shown in Figure 1.9. Iron and steel still account for the largest proportion, although their percentages have been decreasing over the past 25 years. However, the volume of high-grade steel sheets, such as high-strength steels with excellent crashworthiness, and coated steel sheets with excellent anti-corrosion performance is increasing. Iron and steel are expected to remain in first place for some time to come. On the other hand, the use of aluminum alloys to make cylinder blocks, wheels, and other parts is rapidly increasing due to the demand for lighter vehicles. Aluminum alloy sheets have been applied to panels like the engine hood in recent years. This trend is expected to continue in the future.

Future Direction of Automotive Materials

Materials have contributed to meeting the changing requirements for vehicles over the years. In the future, contributions of material technologies will continue to be needed in two principal fields, the environment and safety.

The projected future direction of related technologies in each field is discussed in the following sections.

Environmental Viewpoint

Issues that are important for environmental protection include reducing exhaust emissions, using clean energy, reducing pollutants, improving fuel economy, and recycling, among others. New material technologies are needed to address these issues, as shown in Figure 1.10.

A diesel engine achieves better fuel economy than a gasoline engine. A direct-injection engine makes it possible to improve fuel economy further by means of lean burning. However, these two types of engine need an after-treatment system for the emission gas. A particulate filter is needed for diesel engines and an NO_x catalyst for direct-injection engines. There are strong needs for the development of high-power batteries and high-performance magnets for electric motors, which will be used on vehicles equipped with a hybrid engine or with a fuel cell that is expected to be the ultimate vehicle power source with no harmful exhaust gas. Moreover, development of new materials for fuel cells is also needed.

Vehicle weight savings are very effective in improving fuel economy, because the vehicle weight accounts for 30% of the total fuel consumption loss. Applying higher strength steels to body structural parts and aluminum alloys and/or plastics to body panels will make a large contribution to reducing vehicle weight. Moreover, applying higher strength materials to powertrain components will also make a large contribution to reducing the size and weight of these parts.

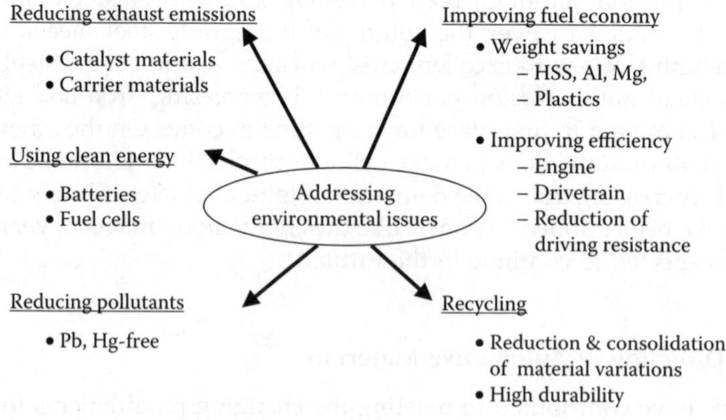

FIGURE 1.10
Important issues for environmental protection.

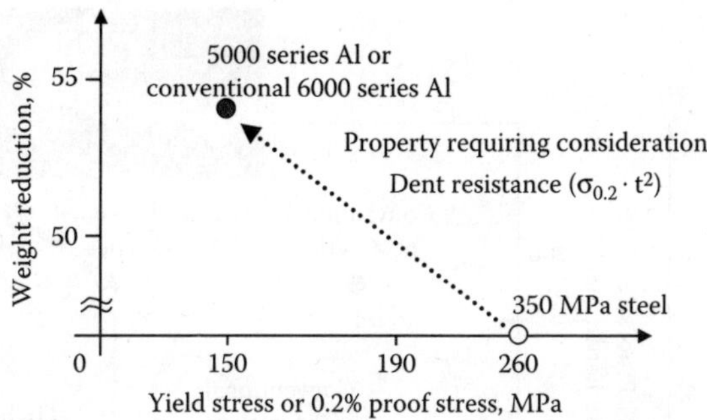

FIGURE 1.11
Reduction of outer panel weight by substituting aluminum for steel sheet.

Figure 1.11 shows an example of the use of aluminum sheet for outer body panels. Dent resistance is one property that must be taken into consideration when lightening outer panels. Substituting aluminum for steel sheet would make it possible to reduce the panel weight by more than 50%.

However, formability is an important factor in the extensive application of aluminum sheets to body panels. The property of dent resistance, needed for outer panels, is determined by 0.2% yield strength, as shown by the following relationship:

$$\text{Dent resistance} \propto (\sigma_{0.2} \times t^2) \tag{1.1}$$

where

$\sigma_{0.2} = 0.2\%$ yield strength
t = sheet thickness

6000 series aluminum alloys have higher yield strengths than 5000 series alloys, and 6000 series sheet provides correspondingly larger weight savings. However, 6000 series aluminum alloys have poorer formability than 5000 series alloys, which limits the application of 6000 series alloys to body panels. The trunk lid requires a sheet with good formability, so 5000 series alloys are generally used. However, newly developed 6000 series aluminum alloys could be applied to the trunk lid, because, although yield strength is lower during the forming process, it increases after paint baking, as shown in Figure 1.12. Developments in aluminum alloy body panels and sheet are discussed in more detail by Takashi Inaba in Chapter 2.

Meanwhile, different approaches are being taken to lighten vehicles through efforts to redesign the frame structure and panel parts. Audi is producing a vehicle with an all-aluminum body-in-white. In addition to changing the traditional monocoque body structure to a space frame construction, Audi switched the body material from steel to an aluminum alloy. This aluminum space frame structure deserves attention because of its cost-saving potential, depending on the vehicle production volume.

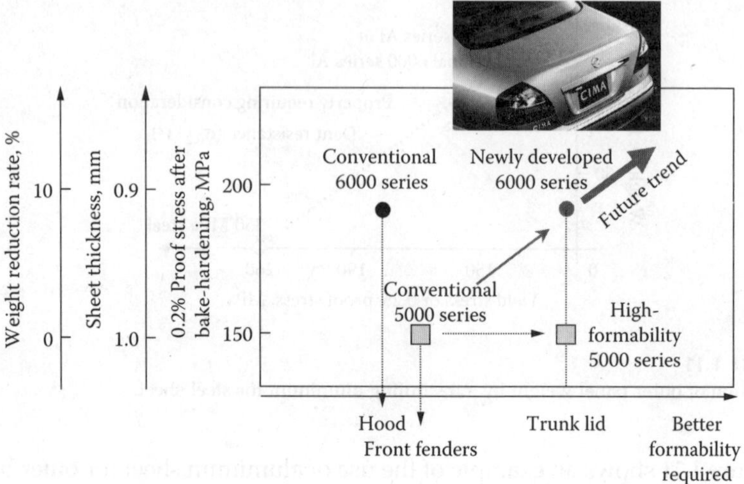

FIGURE 1.12
Trends in aluminum sheet usage for outer panels.

On the other hand, magnesium alloys are being used only in small quantities in the automobile today. However, magnesium alloys could have a large effect on reducing vehicle weight due to their low density. Therefore, it is hoped that technologies will be developed for applying magnesium alloys to automotive components.

Friction in an engine accounts for 40% of all the fuel consumption loss. There is a need to develop technologies for reducing the friction coefficient and weight of engine components, in particular the valve train and piston-crank systems, in order to contribute to improving fuel economy. Higher wear-resistant materials and surface treatments are needed for reducing load stress by lightening the weight of components and reducing the contact area.

Safety Viewpoint

Material technologies are also expected to contribute to improving crashworthiness. In order to achieve a safe car body in the event of a collision, deformation of the cabin structure should be minimized to protect the occupants, and the collision energy should be absorbed in a short deformation length within the crushable zones, as shown in Figure 1.13.

However, the reaction force generally exceeds an appropriate level when a material with higher strength is applied to an energy-absorbing location.

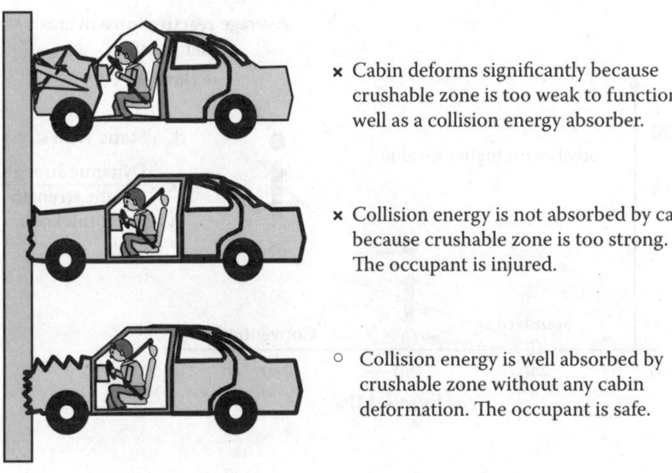

FIGURE 1.13
Concept of crash safety.

Consequently, new structures and materials are required for building the ideal car body that can absorb the collision energy in a short span and with a constant reaction force.

To meet the requirements for improved safety, thicker steel sheets or additional reinforcements are usually applied, which leads to a heavier body-in-white. Therefore, it is necessary to improve crash safety while at the same time lightening vehicles for better environmental performance.

From the viewpoint of materials, both dynamic strength and static strength are important in designing parts for greater crash safety. As defined in

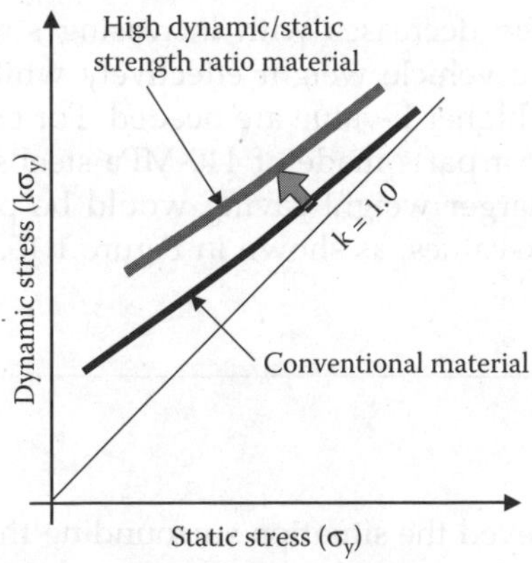

FIGURE 1.14
Relationship between static strength and dynamic strength.

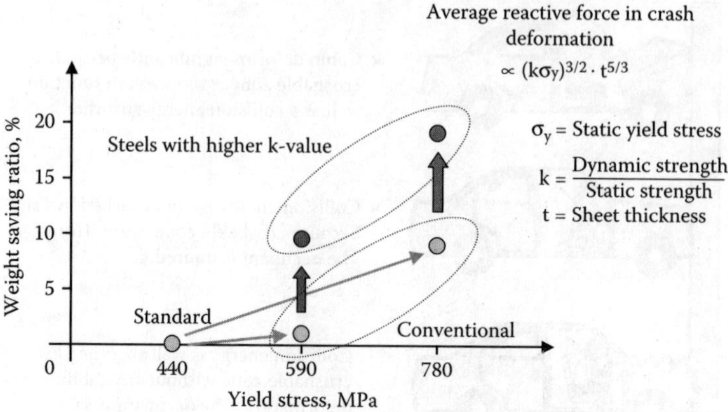

FIGURE 1.15
Part weight reductions achieved by using high-strength steel with a higher k-value.

Equation 1.2, the average reactive force of a rectangular tube with a hat-shaped cross section is related to the k-value, i.e., the dynamic/static ratio of yield strength[5]:

$$\text{Average reactive force in crash deformation} \propto (k\sigma_y)^{3/2} \times t^{5/3} \qquad (1.2)$$

where

k = dynamic yield strength/static yield strength
σ_y = static yield strength
t = sheet thickness

In general, the k-value decreases with increasing strength, as shown in Figure 1.14. To reduce vehicle weight effectively while improving safety, new materials with a higher k-value are needed. For example, substituting higher strength steel for parts made of 440-MPa steel sheet can reduce the weight, but a much larger weight saving would be possible by applying steels having higher k-values, as shown in Figure 1.15.

Summary

This chapter has surveyed the situation surrounding the automotive industry, including the requirements for environmental friendliness and crash safety, from the viewpoint of the harmonious coexistence of human beings, nature, and vehicles. The discussion of the future direction of material technologies has shown that various improvements can be attained by improving material characteristics.

However, in order to apply a new material to a vehicle, cost competitiveness and the availability of a global supply both need to be ensured. At the same time, peripheral technical issues such as forming and joining technologies and environmental performance should also be addressed. Regarding the cost of materials, one guideline for future material selection is likely to be a specified level of cost performance from the customer's viewpoint. Moreover, in order to overcome these technical issues, simultaneous or concurrent engineering by materials suppliers, parts suppliers, and car manufacturers, or among car manufacturers, is becoming more important than ever before.

References

1. Japan Automobile Manufacturers Association, Inc. (JAMA): Japanese Automotive Industry, 2001 (in Japanese).
2. IEA/OECD: *World Energy Outlook*, 1998.
3. JAMA Web site: http://www.jama.or.jp.
4. Source: Japan Automobile Research Institute, Inc.
5. Aya, N., and K. Takahashi, *Energy Absorbing Characteristics of Body Structures (Part 1)*, JSAE, Vol. 7, 60, 1974 (in Japanese).

2

Automobile Aluminum Sheet

Takashi Inaba

CONTENTS

Introduction

In recent years, environmental improvement and safety have become very important for the automobile industry. Environmental improvement and safety features lead to increases in car body weight. To reduce weight, therefore, it is necessary to select optimum materials such as aluminum alloys.

Figure 2.1 shows the plan to reduce CO_2 emissions in Europe.[1] European automobile manufacturers have to achieve an average CO_2 emission target of 140 g/km for their fleet of new cars to be sold in 2008.[2,3] Japanese automobile manufacturers have to achieve the same target by 2009. In North America and Japan,[1] automobile manufacturers also have to achieve fuel consumption regulation targets. For these reasons, aluminum alloys are essential to reduce the weight of car bodies.

This chapter provides general information on how aluminum body panels are used in Europe, North America, and Japan. The promotion of increased

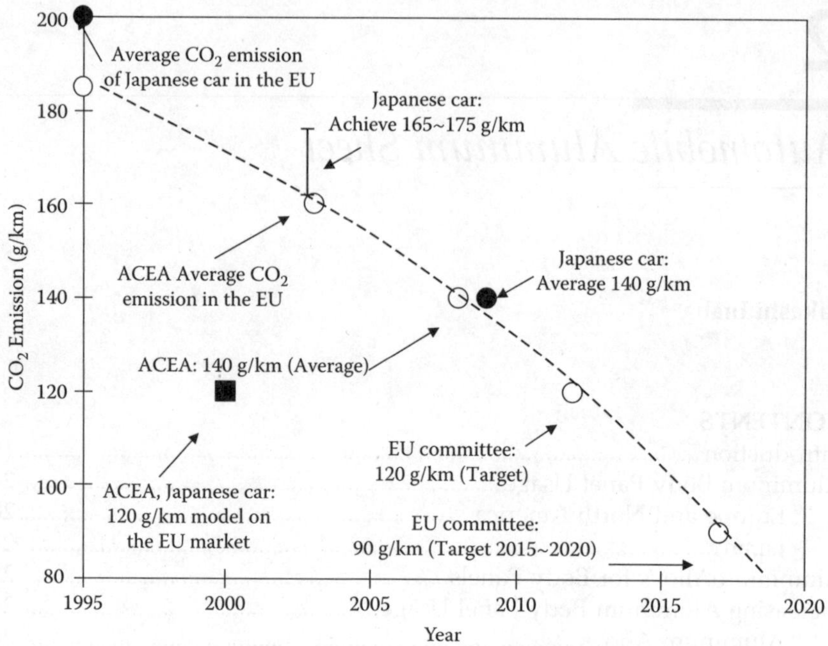

FIGURE 2.1
Plan to reduce CO_2 emissions in Europe.

aluminum body panel use and possible recycling opportunities are also discussed.

Aluminum Body Panel Usage

Europe and North America

Aluminum body panels are used for luxury cars, popular cars, and full-size cars in Europe and North America, as shown in Table 2.1. The automobile manufacturers are mainly using only aluminum hoods except for special cases where they are making all-aluminum cars. The use of aluminum hoods is effective for both weight reduction and improved function as a hang-on part. The adoption of aluminum panels is limited at present by the complexity of the panel shapes, but the use of aluminum panels will increase substantially in the future as automobile manufacturers strive to achieve the CO_2 emission targets in Europe, and the fuel consumption regulation targets in North America.

TABLE 2.1

Examples of Adoption of Aluminum Panels in Europe and North America

Europe	Benz S-class	Hood
	Benz E-class	Hood, fender, deck-lid
	Audi A8,A2	All-aluminum car
	Audi A6	Hood
	Volvo S60	Hood
	Volvo S70	Backdoor
	VW Lupo	All-aluminum car
	Renault Laguna	Hood
	Peugeot 307	Hood
	Citroen C5	Hood
North America	GM Cadillac Seville	Hood
	GM C/K Truck	Hood
	Ford Lincoln	Hood
	Ford Ranger	Hood
	Ford F150	Hood
	Chrysler Prowler	All-aluminum car
	Chrysler Jeep	Hood

Japan

The use of aluminum body parts started with the hood of the Mazda RX-7 in 1985. The Honda NSX all-aluminum car followed in 1990. At first, aluminum body panels were adopted for parts of sport cars in Japan, but recently they have been used for mass-produced cars such as the Nissan and Subaru cars shown in Table 2.2. Aluminum body panels are also used for the compact Copen car produced by Daihatsu.

TABLE 2.2

Examples of Adoption of Aluminum Panels in Japan

	Toyota Soarer	Hood, roof, deck-lid
	Toyota Altezza Gita	Backdoor
	Nissan Cedric	Hood
	Nissan Cima	Hood, deck-lid
Japan	Nissan Skyline	Hood
	Honda S2000	Hood
	Honda Insight	All-aluminum car
	Mazda RX7	Hood
	Mazda Roadster	Hood
	Mitsubishi Lancer Evo	Hood, fender
	Subaru Legacy	Hood
	Subaru Imprezza	Hood
	Daihatsu Copen	Hood, roof, deck-lid

TABLE 2.3

Important Properties Required for Body Panels

Panel	Main Properties
Outer	• High strength after paint baking (YS: 200 MPa at 170°C for 20 min after 2% strain) • Flat hemming property • Surface condition (SS-mark free, anti-orange peel) • Anti-corrosion (anti-filiform corrosion)
Inner	• Deep drawing property • Joining properties (welding, adhesion)

Aluminum Alloys for Body Panels

Automobile body panels consist of a double structure with an outer panel and an inner panel. For the outer panels, higher strength materials are especially required to provide sufficient denting resistance. For the inner panels, higher deep drawing capacity materials are especially required to allow the manufacture of more complex shapes. In other words, different properties are required for the outer and inner panels, as shown in Table 2.3.

Research and development of aluminum body panels began in the 1970s. Aluminum alloys for body panels developed in different ways in Europe, North America, and Japan because of the different requirements of the automobile manufacturers. In Japan, higher formability alloys were required from the automobile manufacturers. Therefore, special 5xxx series Al-Mg alloys, such as AA5022 and AA5023, were developed first. On the other hand, high strength alloys after paint baking were required in Europe and North America. Consequently, 2xxx series Al-Cu-Mg alloys, such as AA2036, and 6xxx series Al-Mg-Si-(Cu) alloys, such as AA6016, AA6111, and AA6022, were developed. The mechanism of paint bake-hardening of 6xxx series alloys is due to precipitation hardening of Mg_2Si or a Cu-containing derivative. Figure 2.2 shows the transition of aluminum alloys for body panels.

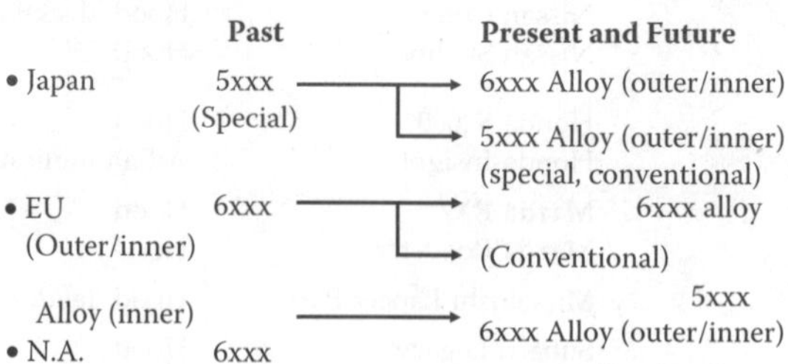

FIGURE 2.2
Transition of aluminum alloys for body panels.

TABLE 2.4

Chemical Composition of Aluminum Alloys for Body Panels (wt%)

AA No.	Si	Fe	Cu	Mn	Mg	Remark
6016	1.0/1.5	<0.50	<0.20	<0.20	0.25/0.60	Outer/Inner, EU (Kobe: KS6K21)
6022	0.8/1.5	0.05/0.20	0.01/0.11	0.02/0.10	0.45/0.70	Outer/Inner, USA (Kobe: KS6K21)
6111	0.07/1.1	<0.40	0.50/0.90	0.15/0.45	0.50/1.0	Outer/Inner, USA (Kobe: KS6K31)
5022	<0.25	<0.40	0.20/0.50	<0.10	3.50/4.9	Outer/Inner, Jap. (Kobe: KS5J30)
5023	<0.25	<0.40	0.20/0.50	<0.10	5.0/6.2	Outer/Inner, Jap. (Kobe: KS5J32)
5182	<0.20	<0.35	<0.10	0.20/0.50	4.5/5.0	Inner, EU and Jap.

Recently, similar 6xxx series alloys have been used in Europe, North America, and Japan.

Table 2.4 shows the chemical compositions of aluminum alloys for body panels. AA6016 contains less than 0.2% Cu content, and is used in Europe. AA6111 contains higher Cu content than AA6022. Both alloys are used in North America. Alloys similar to low Cu content AA6016 and AA6022, and high Cu content AA6111 are also used in Japan. KS6K21 and KS6K31 are alloy codes of Kobe Steel, which correspond to AA6016, AA6022, and AA6111 respectively. AA5022 and AA5023 are special Al-Mg alloys produced by using high purity primary aluminum. They contain optimum Cu content, and have high formability and medium strength after paint baking. KS5J30 and KS5J32 are corresponding Kobe Steel alloys. These alloys are still in use for body panels of severe complex shapes in Japan. For inner panels, the conventional 5xxx series AA5182 alloy has been used recently in Europe and Japan.

Table 2.5 shows typical mechanical properties of aluminum alloys for body panels produced by Kobe Steel. KS6K21-1 has high strength, with a yield

TABLE 2.5

Mechanical Properties of Aluminum Alloys for Body Panels (Kobe Steel)

Kobe Alloy	Before Forming			After Baking	Remark
	TS MPa	YS MPa	El. %	YS MPa	
KS6K21-1	240	125	29	200	Outer/Inner
KS6K21-1	250	130	30	165	Inner
KS6K31	275	130	32	165	Inner
KS5J30	275	135	30	155	Outer/Inner
KS5J32	285	135	33	155	Outer/Inner
5182	270	125	29	140	Inner

strength of 200 MPa after paint baking, and is in use for many body panels in Japan. However, the formability of KS6K21-1 is inferior to that of KS5J32. On the other hand, KS5J32 has higher elongation than KS6K21-1.

Increasing Aluminum Body Panel Usage

In order to promote the adoption of aluminum body panels, it is necessary to provide for the potential panel shapes and the low-cost materials required by automobile manufacturers. It is important to improve material properties as well as forming and joining technologies, so as to be able to manufacture suitable body panel shapes. On the other hand, it is necessary to minimize the number of manufacturing processes, and to be able to use recycled aluminum alloys to ensure a low-cost material.

Aluminum Alloys

The important properties required for body panels are as shown in Table 2.3. Especially, it is necessary to improve the formability to enable, for example, hem flanging and stretch-forming for outer panels, and deep drawing for inner panels. It is important to be able to decrease strength before forming, and then redevelop high strength after paint baking under conventional baking conditions.

Figure 2.3 shows a study of the bake-hardening properties of 6xxx alloys after pre-aging.[4] The specimens are solution heat treated at a high temperature of 530°C and then water quenched, a conventional manufacturing process for aluminum body panels. Pre-aging is conducted at 50°C to 100°C immediately after water quenching. After one week at room temperature, the specimens are then heat treated using several different baking conditions.

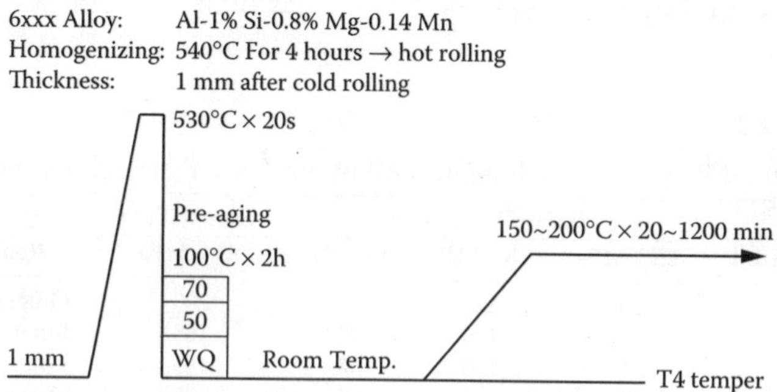

FIGURE 2.3
Study of the bake-hardening properties of 6xxx alloys after pre-aging.

The pre-aged specimens have high strength after paint baking at low temperature for a short time compared with more conventional specimens. The improved bake-hardening properties are caused by fine precipitates of $\beta' - Mg_2Si$. This study is important in indicating how to improve the material properties.

Forming Technology

It is not easy to promote the adoption of aluminum body panels just by improving the material properties. It is also important to provide optimum forming technologies for manufacturing the aluminum body panels. For example, tooling and forming conditions both need to be optimized. In addition, many kinds of forming technologies, such as hydro-forming, hot-forming, and extreme cold-forming, need to be studied. Kobe Steel is investigating the optimization of tool and forming conditions using practical pressing studies and finite element (FEM) analysis.

Figure 2.4 shows the 1400-ton Kobe Steel test press for manufacturing aluminum body panels. Useful data for aluminum body panels compared with conventional steel panels can be achieved by using direct experimental pressing studies.[5,6]

1400-ton hydraulic press

Optimization of tool and forming

FIGURE 2.4
1400-ton Kobe Steel test press for manufacturing aluminum body panels.

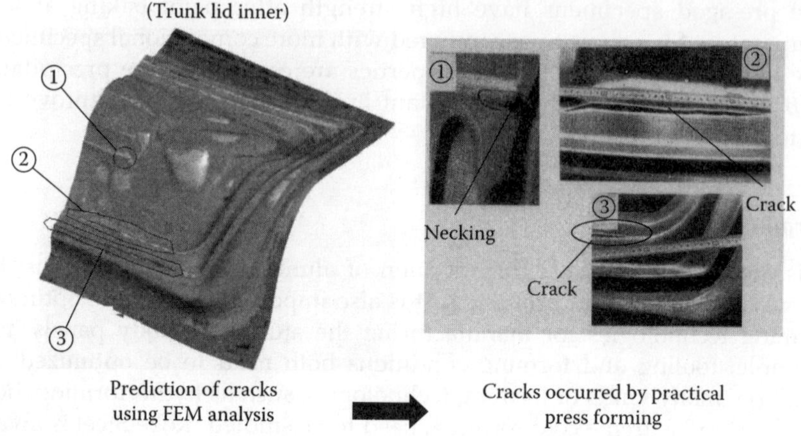

FIGURE 2.5
Relationship between finite element analysis and practical press forming.

On the other hand, Figure 2.5 shows the relationship between finite element analysis and practical press forming.[7] The prediction of cracks using finite element analysis corresponds with the results of cracks occurring during experimental press forming. The precision of finite element analysis will improve with increased applications, and this will play an increasing role in promoting adoption of aluminum panels.

Recycling

Aluminum alloys have excellent recycling properties. It is well-known that used aluminum beverage cans can be returned to new beverage cans. In Japan in 2001, the recovery ratio of used aluminum beverage cans was 83%, with a can-to-can ratio of 68%, the rest being used mainly for castings. Recycling of aluminum alloys is useful for reducing the cost of the aluminum material, and leads to improved life-cycle assessment. Therefore, the reuse of aluminum alloy body panels needs to be studied. In the case of aluminum press scrap, aluminum manufacturers can reproduce the same alloy sheets. However, in the case of aluminum scrap from a scrapped car, it is not easy to recover the same alloy sheet, because of the mixing of different alloys, such as 6xxx, 5xxx, and Al-Si series alloys, and different metals, such as aluminum and steel. Therefore, aluminum manufacturers have to work on alloy designs suitable for recycling and construction of a viable recycling system. The final target will be car-to-car.

Figure 2.6 shows the effect of using an aluminum Audi ASF car on saving energy. Energy saving from an all-aluminum car will be excellent compared with conventional steel cars, with the introduction of recycled aluminum alloys.

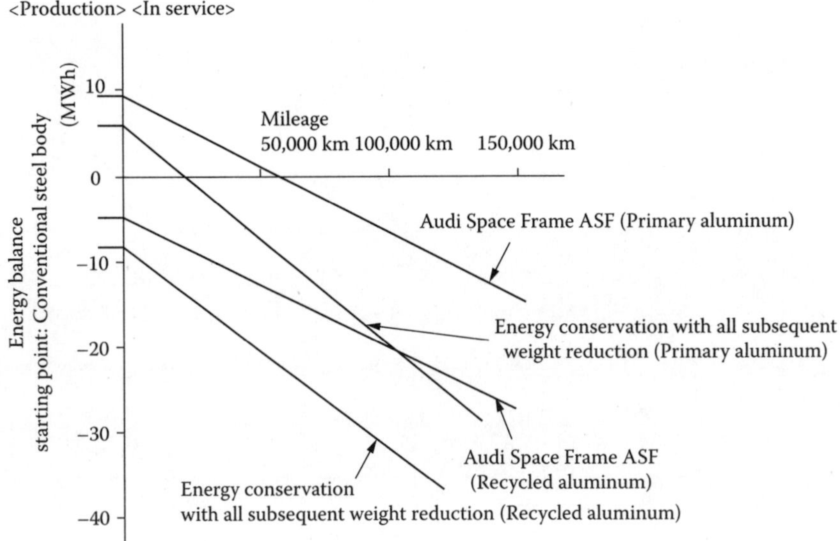

FIGURE 2.6
Effect of using an aluminium Audi ASF car on saving energy.

Summary

The use of aluminum body panels and recycled aluminum alloys leads to weight reduction in car bodies. Therefore, the promotion of the adoption of aluminum body panels is very important in both automobile and aluminum manufacturing industries.

References

1. Minato, K., *Journal of Society of Automotive Engineers of Japan*, 54–9, 2000, 11.
2. Winterkorn, Martin et al., *ATZ*, Vol. 101, 24, 1999.
3. Leitermann, Wulf et al., *Sonderausgabe von ATZ und MTZ (Audi A2)*, 68.
4. Sakurai, T., *The '87 conference of Japan Institute of Light Metals*, 185.
5. Noda, K., *The '97 conference of Japan Institute of Light Metals*, 167.
6. Yoshida, M., *The '89 conference of Japan Institute of Light Metals*, 159.
7. Konishi, H., *The proceedings of the 1999 Japanese Spring Conference for the Technology of Plasticity*, 347.

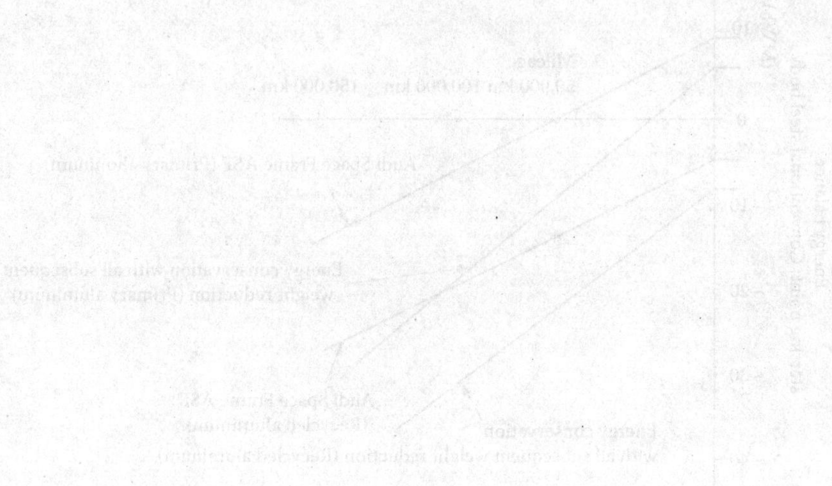

3

Plastic Technology for Automotive Modules

Kazuhisa Toh

CONTENTS

Introduction

Modularization in the automotive industry is a production method that regards plural parts as a single part by consolidating them into a single functional unit. This chapter describes trends in automotive modularization and the associated use of plastic materials and molding processes for module carriers.

European automotive makers have been adopting modular parts mainly for reasons of cost saving since the late 1980s. In Japan, this kind of modularization has not been developed because of insufficient cost benefit. Recently, some Japanese automotive makers have begun to adopt modular parts for cost and weight savings by applying better materials and processes to module carriers. High-performance materials such as sheet molding compound (SMC) or glass mat reinforced polypropylene (PP) are normally used for module carriers because it is necessary to sustain several surrounding parts. Instead of these materials, new materials and processes have been developed and have achieved dramatic cost and weight savings in the resulting modular parts. This chapter also describes further new technologies, such as nanocomposites, which are expected to be used for future module carriers.

Modularization Methods

The subassembly in the body assembly line has been used in Japan since the 1970s to shorten the length of the main line. Recently, "function integrated modules" were introduced to reduce the cost and weight of parts by consolidating them and integrating their functions.

Module Carrier Requirements

Module carriers are the foundation for assembling surrounding plural parts. Mechanical properties such as good strength, stiffness, impact strength, durability, dimensional stability, etc., are all required not just in the module carrier itself, but also after assembly with the surrounding parts. High-performance materials are, therefore, needed to manufacture module carriers.

Development Trends

Sheet molding compound and glass mat reinforced polypropylene have been the main materials used for module carriers since the late 1980s. Both materials are reinforced by glass fibers, have high strength and good dimensional stability, and can meet the requirements for module carriers. However, a semi-finishing product step is required after compression molding, and design flexibility is not so high. Moreover, sheet molding compound cannot easily be recycled because of the presence of thermosetting resin. In order to improve these characteristics, new materials and processes have been developed, as discussed below.

Plastic/Steel Hybrid

In late 1997, a European automotive maker introduced a plastic/steel hybrid structure for a front-end module carrier consisting of injection molded polyamide (PA) and steel reinforcement. The strength of an injection-molded part is normally lower than that of a compression-molded part. In this case, however, the steel part of the component was effective in enhancing the strength. The difference between a hybrid part and a conventional compression-molded part is the need for a semi-finishing product step. The hybrid part does not need post-finish treatment because of the injection molding. According to a material supplier, hybrid structures can reduce cost and weight by 10% each, compared with compression-molded parts.

Long Fiber Thermoplastic (LFT)

A European automotive maker introduced long fiber thermoplastics technology in 1998. First, an intermediate product consisting of polypropylene and 25–80mm glass fibers in length is produced with an extruder. Then, the intermediate product is charged into a molding die and compression molded. This material can be used with a complex design and is also easily recyclable compared to conventional glass mat reinforced polypropylene.[1]

Stamping Mold

The melted resin from the extruder head is directly charged into the die. Then, the material is compression molded. The molded part has low warpage and high-design flexibility. A front-end carrier using this technology is 30% lighter than a conventional steel part. In 2001, a Japanese automotive maker introduced this technology.[2]

Injection-Molded Polypropylene Reinforced by Long Glass Fibers

Injection-molded module carriers have been developed since the early 1990s. In 2002, a Japanese automotive maker introduced injection-molded carriers without steel reinforcement. This system has then been applied to a front-end module and door module carriers for a worldwide series of production cars.

A semi-finishing product step is not necessary for an injection-molded part. In addition, injection-molded parts have a high level of design flexibility. However, injection-molded strength is lower than for conventional glass mat reinforced polypropylene because the length of the glass fibers is shortened during the injection-molding process. The length of the glass fibers in a conventional injection-molded part is reduced to less than 2 mm from an initial length of 10 mm. The length of the glass fibers needs to be more than 4 mm for a high-strength part. Accordingly, new materials and processes have been developed to inhibit breakage of the glass fibers during injection molding.[3] This new technology is discussed below.

New Materials

In conventional injection molding a high-molecular weight, high-viscosity resin is generally used in order to improve the molded-part strength. However, in recent technological developments, a super-low viscosity resin is used instead, in order to maintain the length of glass fibers by reducing shear forces on the glass fibers during the molding process. As a result, much longer glass fibers are maintained in the molded part and its strength is increased.

New Processes

New mixing screws for large-scale injection molding machines have also been developed in order to inhibit the breakage of the glass fibers by reducing the shear force on the glass fibers. The mechanical properties of the resulting injection-molded parts are much superior to those of conventional injection-molded parts and equivalent to those of the conventional compression-molded parts (Table 3.1).

Applications and Benefits

About 20 parts have been consolidated into one part for a front-end module carrier by using the newly developed injection-molding technology described above. The injection-molded carrier is then 25% cheaper and 18% lighter than using a conventional steel part. Moreover, the same technology and material has also been applied to a door module carrier. This carrier is then 20% cheaper and 2.3 kg lighter than using a conventional steel part by integrating plural parts into the carrier plate.

TABLE 3.1

Mechanical Properties of Injection-Molded and Compression-Molded Parts

	Conventional Compression-Molding	Conventional Injection-Molding	Developmental Injection-Molding
Flexural modulus (GPa)	5.3	5.1	5.2
Flexural strength (MPa)	139	115	126
Izod notched impact strength (kJ/m²)	29.9	9.3	32.5

Future Technology

Modularization will be further promoted and expanded by improving mechanical properties, surface quality, recyclability, etc., of the module carrier. The potential technology that will be able to meet these requirements is discussed below.

Future Materials

Nanocomposites

The mineral clay is attracting attention as a reinforcement filler for injection-molded polyamide parts, enhancing the part strength considerably more than conventional fillers such as glass fibers. The aspect ratio of the nanoscale clay particles vary from several hundreds to several thousands—very high compared to conventional glass fiber filler. The resulting injection-molded nanocomposite polyamide materials show unique characteristics. Their strength and stiffness at high temperature are dramatically improved at the same time as maintaining the chemical resistance, paintability, surface appearance, etc., that are important polyamide material characteristics.[4] Polyamide nanocomposites are expected to be applied to module carriers for a range of visible parts. Polypropylene nanocomposites are also being researched. It is reported that 4.5% polypropylene/clay nanocomposites have strengths equivalent to 20% talc-reinforced polypropylene.[5]

High-Strength Plastic Reinforced by Liquid Crystal Polymers (LCP)[6]

Thermoplastic resins are generally recyclable or can be reused for similar applications. However, resins reinforced by glass fibers lose their performance after the granulating process in recycling because of the damage sustained by the reinforcing fibers. An important development target is to increase the strength of the polymer composite and make it even easier to recycle. Composites have been prepared using a generic twin-screw extruder to blend liquid crystal polymers with thermoplastic resin, with extrusion conditions such as shear rate at the die set to cause fibril formation. The liquid crystal polymer, polypropylene, and compatibilizer are mixed initially in pellet form, and then extruded in a composite film with the polypropylene reinforced by liquid crystal polymer fibrils in the extrusion direction. These films are pre-heated and laminated to produce a moldable blank with good moldability, and compression-molded samples exhibit good mechanical properties. The stiffness of a bumper beam prototype made of polypropylene/liquid crystal polymer is equivalent to a conventional glass mat reinforced polypropylene material. Figure 3.1 shows load-displacement curves for equivalent polypropylene/liquid crystal polymer and the glass mat reinforced polypropylene bumper beams.

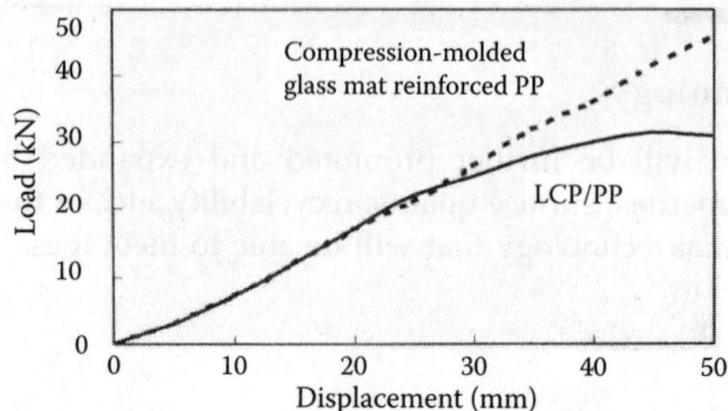

FIGURE 3.1
Load-displacement curves for injection-molded polypropylene/liquid crystal polymer and compression-molded glass mat reinforced polypropylene composites.

The recyclability of the liquid crystal polymer composites is shown in Figure 3.2. Even after being recycled ten times, the tensile strength is equivalent to that of the virgin material. At present, the price of liquid crystal polymer is about ten times that of glass fiber; however, cost reductions are expected to be achieved by expanding applications to the automotive industry.

Future Processing Techniques

Microcellular plastic is manufactured by a new molding process, with improved mechanical and thermal properties compared with conventional foam plastics. The key characteristics of the microcellular plastic material[7] are the cell size, in the range 0.1 to 10 μm, and the cell number density, in the range 10^9 to 10^{15} cells/cm^3. A saturated polymer gas solution is brought

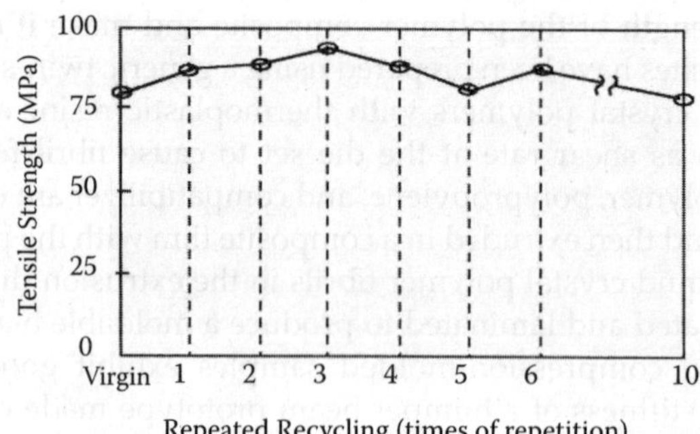

FIGURE 3.2
Tensile strength versus recycling time for a polypropylene/liquid crystal polymer composite.

into a thermodynamically unstable state, where decompression and/or heating instantaneously reduce the gas solubility and leads to formation of the microcells. Carbon dioxide (CO_2) or nitrogen (N_2) gas is usually used to reduce environmental dangers. The resulting characteristics of the foamed plastics:[8]

Low warpage

Low clamping force on the molding machine (i.e., low injection pressure)

Easy to mold large parts

Short molding cycle time

At present, this new process has some technical issues such as poor surface appearance caused by flow marks. However, in the future, the new manufacturing technology can be expected to be used to mold large-scale parts with complex designs, at low cost and weight. These characteristics will be very useful for future module parts.

Summary

This chapter has described materials and processes for a module carrier. Other techniques, such as bonding and computer-aided design, are also very important for automotive modules. By combining the different techniques carefully, plastic technology will contribute strongly toward the development of future automotive modules.

References

1. Nikkei Mechanical, 1998, Vol. 12, No. 531, 17.
2. Tachibana, M., et al., "Plastic Bolster (Radiator Core Support)," *Nissan Technical Review*, No. 50, 47–50, 2002.
3. Tochioka, T., et al., "Development of module carriers by injection molding with glass fiber reinforced polypropylene," *JSAE*, No. 31-02, 5–8, 2002.
4. Ogami, A., *JSPP*, No l. 14, No. 4, 217–221, 2002.
5. SPE Automotive TPO Global Conference, 1999.
6. Sasaki, K., et al., "High-strength Plastic Reinforcement by Liquid Crystal Polymers," *FSITA '96*, M0 6.04, 1996.
7. Baldwin, D., et al., Suh, N. P., Polym. Proc., 1994, 6, 187, 245, 1994.
8. Tsuda, F., *JSPP*, Vol. B., No. 2, 83–87, 2001.

Section 2

Functional Materials

Functional materials, those materials that have electronic, optical, or magnetic properties, are an essential part of the modern automobile. Functional materials are not only deployed in the primary structures and drive systems but also in the safety systems and in telematic and entertainment systems, which often provide the key product differentiation features and can yield a high added value for the manufacturers. This section outlines the application of functional materials in five representative areas of automotive technology:

- Combustion sensors
- Controlled rheology fluids for mechanical linkages
- Engineered crash structures
- Engine control and drive sensor electronics
- Smart structures

The authors review the current state of the art and project future application areas where functional materials and smart technologies will make significant impact on automotive design and manufacture.

4

Automotive Catalysts

Michael Bowker

CONTENTS

Introduction

Pollution is not a new phenomenon, and probably existed in the most ancient of times. Indeed, legislation was enacted many centuries ago in different parts of the world. For instance, in England, in the thirteenth century, King Edward I issued edicts aimed at preventing the sulphurous local pollution in London due to coal burning. Most pollution events in the world since then have been associated with the use of fossil fuels, and this is still the source of many problems today. Types of pollution that are topical are given in Table 4.1. Much pollution in recent times has come from the ever-expanding use of cars, and the amelioration of this pollution problem is the subject of this chapter.

The Development of Automotive Catalysts

Pollution was not perceived as a problem during the first few decades of the use of automobiles, due to their relative rarity and, therefore, the low total pollution burden on the atmosphere, notwithstanding the fact that early vehicles

TABLE 4.1

Some Major Environmental Stresses

Environmental Symptom	Likely Cause
Tropospheric pollution, smogs, etc.	Fossil fuel overconsumption
Acid rain	Fossil fuel overconsumption
Global warming	Fossil fuel overconsumption
Ozone hole	Chlorofluoro–carbon emissions
Decreased male fertility	Groundwater pollution

were inefficient. However, with the buildup of car ownership in the world, the pollution load increased until the effect became severe and noticeable. This was recognized first in the Los Angeles basin due to a combination of local factors, but most importantly due to the high earning power of the local population and, therefore, the high per capita car ownership. Smogs occurred in the late 1950s and 60s, which were basically due to emissions from cars that interacted together in the presence of intense sunlight to produce photochemical smogs. The major polluting components were nitrogen oxides (NO_x), ozone O_3, and hydrocarbons together with highly damaging partially oxidized products. One example of these—peroxy acetyl nitrate (PAN)—is detrimental to the lungs in ppb concentrations, and is an eye irritant. Many deaths resulted from such smogs, mainly among the elderly, infirm, or those with existing lung problems.

Finally, the California legislature decided that the problem must be solved and that it could only be achieved in a mandatory way. At first, this was designed to restrict carbon monoxide and hydrocarbon emissions. Later, NO_x was included in the legislation. Initially, the U.S. car companies opposed such legislation. However, a catalyst cure for the problem was not only possible, but was achieved and demonstrated for a production model car. A schematic illustration of the form of a car catalyst currently used is given in Figure 4.1. This is a ceramic monolith that is strong, resistant to thermal shock, and the active phases are present on this monolith inside a highly porous, but thin layer of washcoat, which is usually mainly alumina. More extensive reviews of this technology are available.[1]

From this point in time, the legislators were encouraged to push the scientists to ever-greater improvements of catalytic efficiency by increasing the stringency of legislation. The development of this legislation is illustrated in Table 4.2. It is now the case that catalysts, at least for normally aspirated petrol engines, are very efficient. The current and future challenges to automotive pollution removal are outlined below.

Important Factors in Pollution Removal

Light-Off Temperature

Pollutant emission has to meet the strict legislative levels shown in Table 4.2. These limits are set to get tougher. If the pollutants emitted from the engine

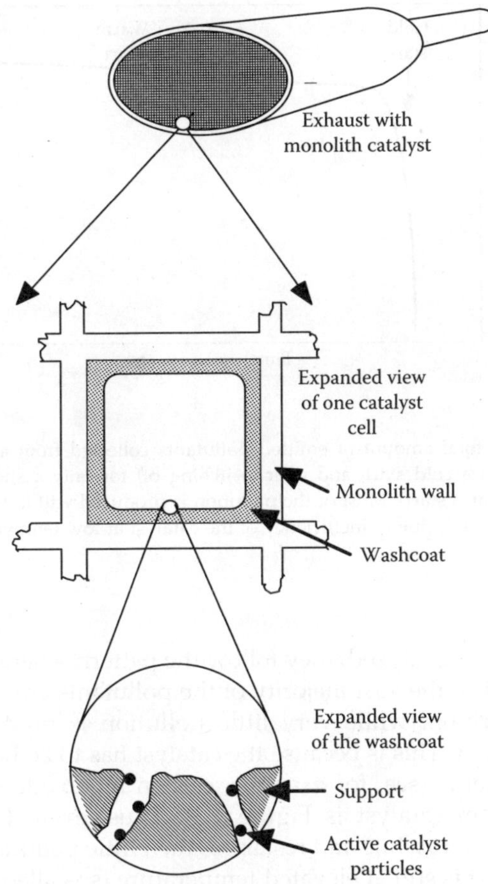

FIGURE 4.1

Schematic diagram of a typical car catalyst, showing the monolith with a washcoat that is impregnated with precious metals and other active species.

TABLE 4.2

Legislated Passenger Car Emissions

Year	Location	Maximum Levels (g/Km)		
		CO	HC	NO_x
1970	United States	14	1.3	-
1975	California	5.4	0.5	1.2
1980	United States	4.2	0.2	1.2
1993	California	2.0	0.2	0.2
1993	European Union	2.7	1.0 (HC and NO_x combined)	
1997	California	2.0	0.05	0.12
1997	European Union	2.3	0.3	0.25
2001	European Union	2.3	0.2	0.15
2005	European Union (Proposed)	1.0	0.1	0.08

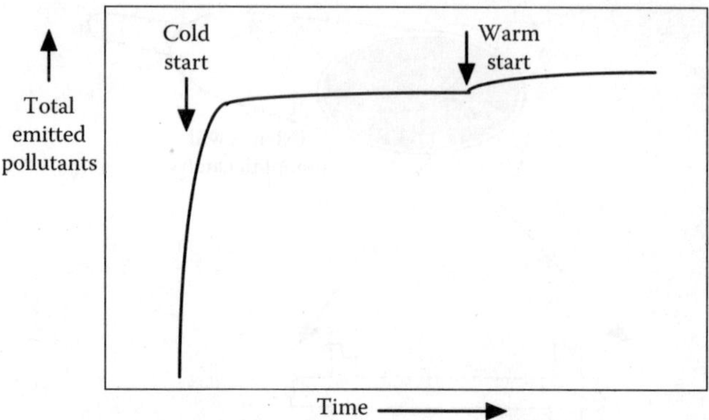

FIGURE 4.2

An illustration of the total amount of emitted pollutants collected from a car exhaust after engine switch-on from a cold start, and after switching off for only a short period of time followed by restart (warm start). Most of the pollution is produced within the first minutes of the engine is switching on, due to inefficiency of the catalyst at low temperatures.

are accumulated and analyzed, they follow the pattern schematically shown in Figure 4.2,[2] that is, the vast majority of the pollutants are emitted shortly after engine switch-on, while very little pollution is produced once the exhaust region is hot. This is because the catalyst has to be hot to be able to convert the pollutants (see, for example, carbon monoxide oxidation by a supported platinum catalyst is Figure 4.3). Thus, even if the engine is switched off for a short while and restarted, then little pollution is produced because the catalyst is still at elevated temperature (so-called "warm start") and is immediately effective upon restart.

Thus, reducing the temperature for light-off to occur is a major target for pollutant reduction, and this requires the development of more active catalytic materials. One way to get light-off to occur more quickly without activity enhancement is to place a catalyst in the engine manifold, close to the source of pollution, since this gets hot more quickly than the normal position farther down the exhaust pipe. This approach is carried out nowadays. Another option being seriously investigated by a number of companies is the use of a plasma discharge prior to the normal catalyst that produces activated species that particularly enhance NO_x destruction. Also, artificial gas heaters in the car exhaust prior to the catalyst have been considered, but these use a significant amount of power from the engine.

There are hopes for more active catalysts that light-off at much lower temperatures than platinum, and one example of this comes from a surprising element, namely gold. Thought previously to be inactive, it has been shown that gold nanoparticles are very active for carbon monoxide oxidation, even at room temperature, when prepared in the correct way.[3]

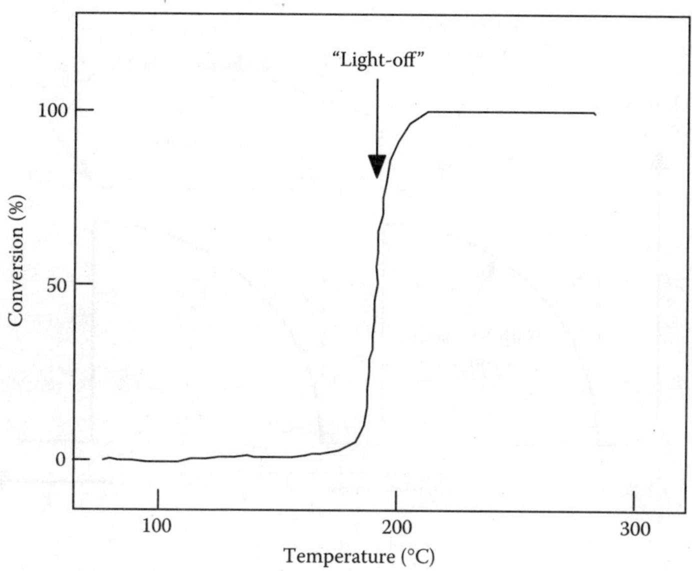

FIGURE 4.3

A plot showing the light-off for carbon monoxide oxidation occurring at about 180°C for a supported platinum catalyst. There is zero conversion below about 150°C.

Lean-Burn Engines

Lean-burn engines have been introduced because they are more fuel-efficient and therefore produce less carbon dioxide burden on the atmosphere, besides being more economically efficient for the user and reducing the rate of loss of fossil fuel stocks. However, although oxidation reactions of hydrocarbons and carbon monoxide occur well under such circumstances, the reduction of NO_x to nitrogen becomes very inefficient. Therefore, new catalytic strategies are needed.

One of these strategies involves the use of a NO_x storage medium in the catalyst, and catalysts based on barium oxide BaO have been developed successfully by Toyota.[4,5] In this situation, when the active metal component is deactivated by becoming saturated with oxygen atoms at the surface, it no longer dissociates NO efficiently. The NO_x is then stored on BaO as the nitrate. Periodically, a reductant (fuel) is injected over the catalyst and this reduces the metal surface so that it becomes active in the direct sense of catalyzing pollutants in the exhaust gas, but it also catalytically decomposes the nitrate and cracks the resulting NO_x to give nitrogen (and oxidized products such as carbon dioxide and water). The effect of the storage medium is shown schematically in Figure 4.4 and results in significantly enhanced NO_x conversion. The details of this catalysis are presented further below.

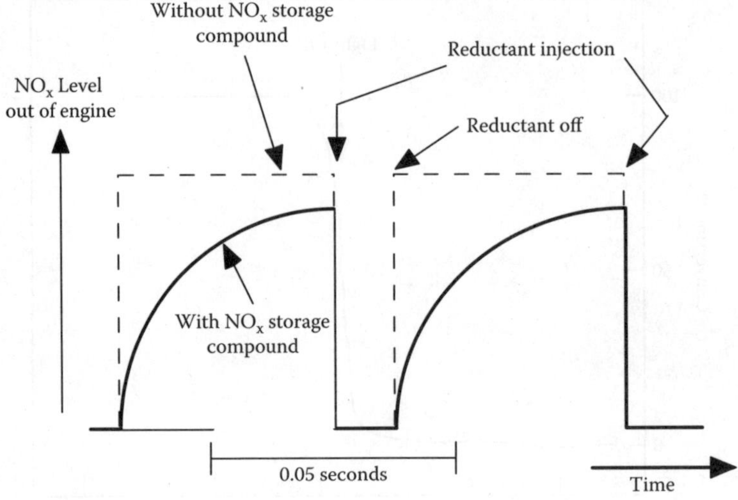

FIGURE 4.4
Schematic illustration of the NO$_x$ conversion process in NSR catalysts. NO$_x$ is stored during the lean operation and is then removed from the catalyst and converted to nitrogen by injection of a short pulse of fuel that chemically reduces metal part of the catalyst that activates it for NO$_x$ destruction.

Diesel

Diesel is a type of lean-burn engine and so some of the above strategies can be used on exhausts of this type. However, a major problem here is that a high level of soot particles are produced from the engine that tend to block and degrade the catalyst since they are not easily removed themselves. Johnson–Matthey in the United Kingdom recently won the MacRobert Award from the Royal Academy of Engineering for their development of the CRT (continuously regenerating trap) particulate removal technology. This is a technology that uses a porous-walled monolith to store the particulates where they are attacked and oxidized by nitrogen dioxide NO$_2$ produced in an initial catalytic oxidation step.[6]

NSR Catalysts

NO$_x$ storage and reduction (NSR) catalysts have a function for binding NO$_x$ during lean running. Figure 4.4 shows schematically the effect of inclusion of barium oxide in such catalysts. As a result of operation with a periodic reduction pulse, significantly more NO$_x$ is converted than would otherwise be the case.

Experiments currently underway at the University of Reading, using hydrogen as the reductant, clearly show the beneficial effect of barium. In Figure 4.5 a pulse of NO is made into a continuous flow of oxygen in helium upon a catalyst that had already been dosed with several pulses of NO. Apart from some displacement of oxygen from the flow, when NO is pulsed there is no net NO uptake on the catalyst and no products are observed. This is because the catalyst is fully saturated with nitrate already under these conditions and the platinum surface is saturated with oxygen atoms, which prevents NO dissociation and barium nitrate decomposition. When a hydrogen pulse is introduced, then there is immediate consumption of the hydrogen with coincident decomposition of the stored NO_x, and a large amount of nitrogen is produced. This is because the platinum is reduced by the hydrogen pulse, adjacent barium nitrate is decomposed to $NO + O_2$, and the platinum cracks the NO in the following way:

$$2NO - 7N_2 + 2O_a$$

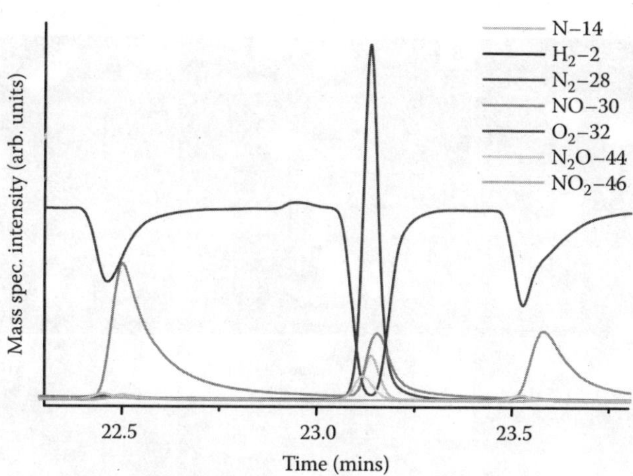

FIGURE 4.5

Showing a pulsed-flow reactor experiment relating to NO_x storage and reduction on a Pt/BaO/Al_2O_3 catalyst. NO (30 amu) is periodically pulsed into a continuous flow of oxygen over the catalyst held at 300°C. The first pulse, shown at 22.5 minutes, is just one of many that the catalyst had previously received, and so it is saturated with nitrate and the whole of the input NO pulse is detected at the reactor outlet. The dip in oxygen (32 amu) here is simply due to displacement from the gas phase by the NO injection, it does not signify reaction. At −23.1 minutes a pulse of hydrogen is injected over the catalyst, which results in reduction of the metal component, and thus considerable extra uptake of oxygen. At the same time a large amount of nitrogen (28 and 14 amu) is evolved due to the catalyzed decomposition of the $Ba(NO_3)_2$ to form BaO. Upon subsequently admitting a pulse of NO at 23.6 minutes, there is uptake of both NO and oxygen (compare the pulse at 23.6 mins with that at 22.5 mins) as decomposed barium nitrate is reformed.

However, once the hydrogen pulse has passed through the catalyst, the platinum reoxidizes again by adsorption of gas phase oxygen. At this point oxygen uptake ceases, the platinum surface has become oxidized, and is unable to decompose the barium nitrate in this condition. However, it can still store NO_x because there are available barium oxide sites surrounding the platinum. Thus, upon admission of the subsequent NO pulses, NO storage and uptake of oxygen is observed. NO storage requires oxygen in the following stoichiometric amount

$$BaO + 2NO + \tfrac{3}{2}\, O_2 - 7\, Ba(NO_3)_2$$

After several further pulses of NO, uptake essentially ceases as the barium oxide again becomes saturated with nitrate.

There are a number of interesting questions related to this catalysis and these concern the atomic and molecular events taking place at the surface of the material. We have begun work at Reading using model catalysts to understand these kinds of reactions. Model catalysts can be made in a variety of ways. For instance, as shown in Figure 4.6, we can make model catalysts

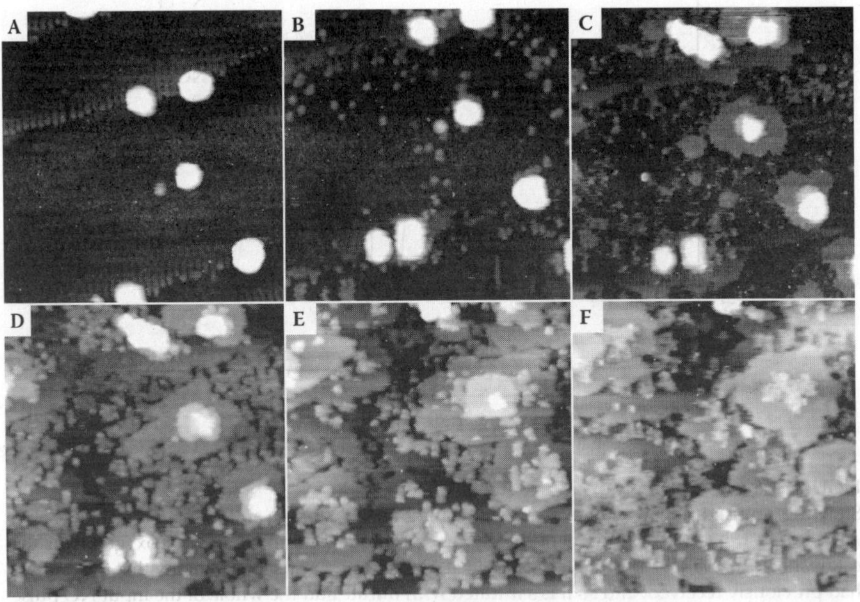

FIGURE 4.6
Showing a series of S1M (scanning tunneling microscopy) images taken from a movie of the effect of gas phase oxygen on the surface structure of a model Pd/TiO_2 catalysts. The Pd is in the form of nanoparticles ~4 nm diameter formed on the surface by metal vapor deposition. Frames A–F represent increasing exposure to oxygen at 673 K, and the size of each image is 50 nm × 50 nm. New layers of titania are grown around the nanoparticles by reaction between oxygen atoms (which form on the Pd and then diffuse off to the adjacent titania) and interstitial Ti^{3+}.

by depositing metal nanoparticles onto an oxidic support. In this case the support is a titanium (110) single crystal and palladium nanoparticles have been formed by MVD (metal vapour deposition).[7] We can view the oxygen storage process on this system as shown in Figure 4.6. Before treatment in oxygen at 670 K, the particles are clearly visible on the surface. However, during oxygen treatment, new layers of titania grow up around the nanoparticles until they eventually completely cover them. At intermediate times, a spillover region is clearly visible.[8] The storage occurs by oxidation of reduced Ti^{3+} species which reside in the bulk of the sample, but that are pulled to the surface during oxidation. This preferentially occurs around the metal particle because oxygen dissociation occurs fast there, but occurs only slowly on the titania itself.

We can also model the NO_x storage process by fabricating inverted catalysts, in this case by depositing barium onto platinum(111) followed by oxidation of the deposited barium. In Figure 4.7a, we see structures resolved at atomic resolution. We believe this is due to a monolayer structure of the barium oxide on the surface, which is clearly not completely homogeneous and probably defected with missing oxygen and barium atoms. We can also make multilayer islands of barium oxide (as shown in Figure 4.7b). When NO and oxygen are introduced, these islands expand due to the formation of barium nitrate, which has a higher volume per barium atom than barium oxide. These preliminary results are part of a bigger programme aimed at a full understanding of the reaction at atomic and molecular level, which should go some way to answering some of the important questions related to the storage phenomenon. This includes such matters as: What is the extent of barium nitrate decomposition? Are the reactions only at the surface of the oxide?

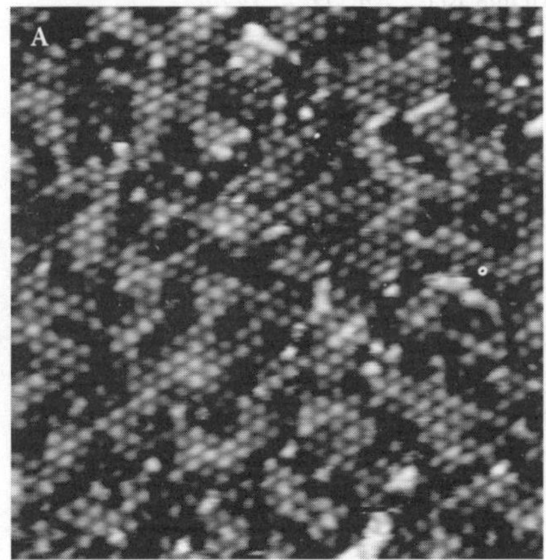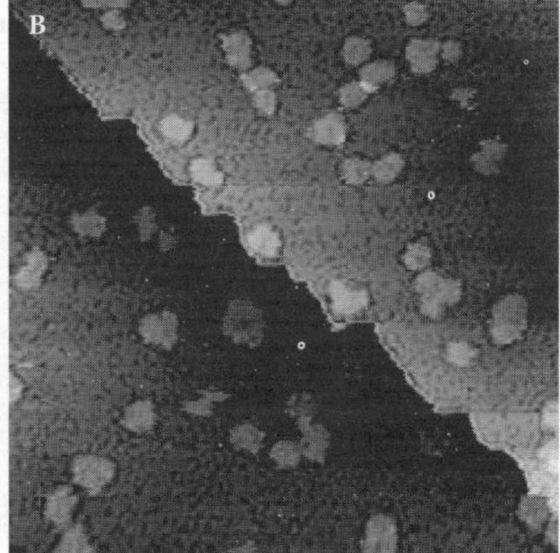

FIGURE 4.7
(a) An atomic resolution image of the Pt(111) surface with a monolayer of BaO dosed on top. Image size 25 nm × 25 nm. (b) An image of the same surface with multilayer islands of BaO.

And what is the nature of the spillover and reverse spillover processes that are responsible for NO_x storage and decomposition?

Summary

This chapter has outlined the development of materials and catalysts for the removal of pollutants from car exhausts. Ways of measuring the catalytic processes involved using time resolved methods are described, as also is the application of a novel, atomically resolving, imaging technique to improve the understanding of the processes involved at the surface of the catalytic materials.

References

1. See, for instance, E. Lax and B. Engler in "Environmental Catalysis," ed. G. Ertl .
2. Bowker, M., "The Basis and Applications of Heterogeneous Catalysis," *Oxford Chemistry Primers*, 1998.
3. Haruta, M., S. Tsubota, T. Kobayashi, H. Kageyama, M. Genet, and B. Delmon, *J. Catal.*, Vol. 144, 175, 1993.
4. Takahashi, N., H. Shinjoh, T. Lijima, T. Suzuki, K. Yamazaki, K. Yokota, H. Suzuki, S. Matsumoto, T. Tyanizawa, T. Tanaka, S. Tateishi, and K. Kasahara, *Catal. Today*, Vol. 27, 63, 1996.
5. Matsumoto, S., *Catal. Today*, Vol. 29, 43, 1996.
6. See, for instance, P.N. Hawker, *PIal. Metals Rev.*, Vol. 39, 2, 1995.
7. Stone, P., S. Poulston, R. Bennett, and M. Bowker, *J. Chem. Soc. Chem. Comm.*, 1369–70, 1998.
8. Bennett, R., P. Stone, and M. Bowker, *Cat. Letts*, Vol. 59, 99–106, 1999.

5

Magnetorheological Fluids

Kevin O'Grady, V. Patel, and S. W. Charles

CONTENTS

Introduction

Controllable fluids that change their mechanical properties under the influence of a remote external influence have been known since the mid to late 1940s. Broadly speaking, two classes of such materials exist: these are electrorheological fluids, which change their rheological behavior under the influence of electric fields, and magnetorheological fluids, which change their rheology under the influence of magnetic fields. These materials are similar to each other in terms of their basic structures in that they are both colloidal dispersions of solid particles in a carrier liquid that is usually a standard hydraulic oil. In the case of electrorheological (ER) fluids, the particles used can be starch or other similar materials, whereas in the case of magnetorheological (Magnetorheological) fluids, the materials used must, of course, be magnetic materials such as particles of iron, cobalt, etc. Both classes of material have a significant number of potential applications in the automotive sector.

Despite the fact that they have been available for many years, the impetus to bring such materials into use has not been great until recently, when issues of noise pollution, both for passengers in automobiles and for those living near to major arterial roads, have become increasingly important. Also, previous generations of materials have perhaps not been suitable for immediate application due to a lack of long-term stability in terms of a number of properties. The application areas are, for the case of dampers, under the seats of heavy vehicles such as lorries, tractors, buses, etc., engine mountings, particularly for vibration and noise suppression in both luxury vehicles, and for example, coaches and buses. It is also possible to design a simple slip clutch or brake that can be used for control of cooling fans and 4-wheel drive differentials. In these cases, the use of a controllable liquid has major advantages in terms of energy conservation. In non-automotive applications, magnetorheological fluids have already found application as generators of a variable resistance in exercise machines. They have also been suggested as an energy-free solution to the problem of maintaining a fire door in the open position that would subsequently close when a fire alarm is sounded.

Historical Developments

Electrorheological Fluids

Electrorheological fluids were developed in the first instance by Winslow in the mid-1940s.[1] Development continued intermittently from that time with a surge in activities in the 1980s and early 1990s with many formulations being proposed both for the fluids themselves and a wide range of devices developed. For a review of these developments, see Scharnhorst and Schelttler-Köhler.[2] The basic structure of the electrorheological fluids were based upon colloidal dispersions where the interaction between the particles in the presence of an electric field derived from the presence of absorbed ions, or more commonly, water molecules on the surface of the colloidal particles themselves. The carrier oils were usually the simple hydraulic oils that are used in normal rheological devices. A wide range of dispersants appropriate to the particular particles and the carrier vehicle were also used.

Modern electrorheological fluids based upon these structures exhibit dynamic yield strengths typically in the range 3–5 kPa for electric fields of the order of 3–5 kV/mm. Unfortunately, given that the basis of the electrorheological effect lies not in a permanent dipole moment but in an induced moment arising from the presence of ionic species or water molecules, this limits the operating temperature of such fluids to the range 10–90°C. Nonetheless, for operation in benign environments, such as seat dampers or retarders on exercise devices, such an operating temperature range is not preclusive.

Magnetorheological Fluids

Magnetorheological fluids were first reported by Rabinov from the National Bureau of Standards in the United States in 1948.[3] The original application envisaged was to produce a device such as a slip clutch.

In the case of magnetorheological fluids, the dispersion consists of fine (but not ultra fine) magnetic particles dispersed in oils, again using conventional dispersants, but also a number of other ingredients as discussed below. Given a correct formulation for such particles, the dynamic strengths are up to 100 kPa for fluids in fields of 2–3 kOe and the temperature range from –40°C to 150°C. These figures must be compared to the much lower 3–5 kPa for electrorheological fluids. This significantly larger yield stress value for magnetorheological fluids derives from the fact that the net polarizing moment per unit volume of a magnetic material derives from its bulk, whereas the electrorheological effect derives only from the surface of the particles and, hence, the effective charge density in the two cases is significantly different.

However, in contrast, the structure of magnetorheological fluids is of necessity more complex. Ideally, it is desirable to use the largest particles possible since the force between them depends upon the product of their volumes. However, in practice, such particles tend to agglomerate irreversibly and normal dispersants are unable to resist the forces of attraction between the particles. There are a number of strategies to overcome this difficulty, one of which is to use dispersants, which are thixotropic agents, and a number of reports of such materials can be found in the literature, e.g., Weiss et al.[4] An alternative strategy is to incorporate thixotropic agents into the dispersion such as clay particles, which prevent the magnetic particles coming into close contact and, hence, prevent irreversible changes in the microstructure of the material.

Structure of Magnetorheological Fluids

Small magnetic particles in the nanosize range (d < 25 nm) are too small to support a normal magnetic domain and, hence, exist in a single domain state effectively behaving as tiny permanent bar magnets. The origin of this effect and the critical size for this behavior varies from material to material, but is generally well understood.[5] Larger magnetic particles typically of dimension 1–100 μm contain a number of magnetic domains that are oriented in such

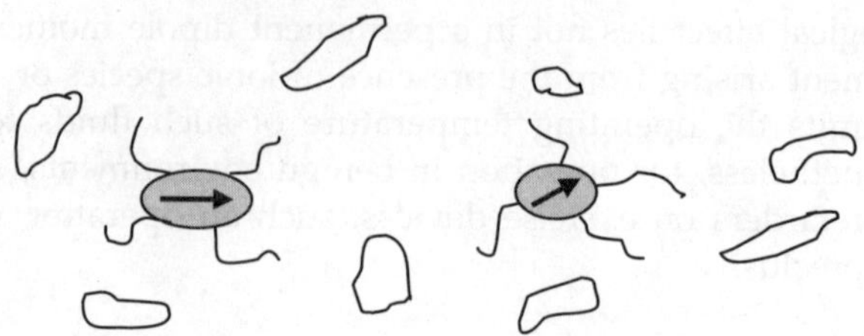

FIGURE 5.1
Dispersed magnetic particles each containing a single magnetic domain.

a way that the external field generated by the particle is minimized. Such an arrangement of domains is termed flux closure and means that such particles have no net magnetic moment in a similar manner to bulk materials. These two different types of magnetic particle are shown schematically in Figures 5.1 and 5.2.

In the presence of a magnetic field, single domain magnetic nanoparticles will experience a force of attraction and, therefore, will come together leading to a magnetorheological effect as large scale structures are formed. However, due to the permanent nature of the magnetic dipole, such particles will tend to adhere together and normal dispersants will not be able to prevent such irreversible agglomeration taking place. In order to prevent irreversible agglomeration, much smaller particles must be used, although in this case the energy of interaction between the particles given by Equation 5.1 below reduces significantly due to the fact that it derives from the volume of material and, therefore, the magnetorheological effect resulting is significantly smaller. Such materials have been developed some years ago but have not found wide application.

As can be seen below the energy of interaction, E_i (in cgs units) is simply given by the product of the magnetic moments μ_1 and μ_2 of the particles and follows an inverse cube law. Given that the magnetic moment of a material is determined by the product of the saturation magnetization M_s and the

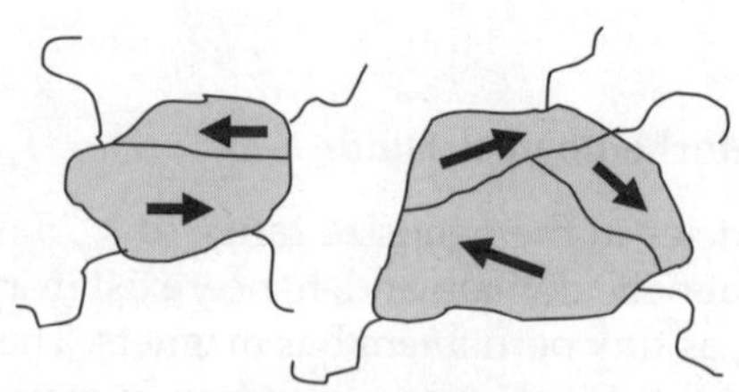

FIGURE 5.2
Dispersed multidomain magnetic particles exhibiting flux closure configurations.

particle volume V,

$$E_i = \frac{\mu_1 \mu_2}{r^3} \tag{5.1}$$

$$E_i = \frac{M_s^2 V_1 V_2}{r^3} = \frac{M_s^2 \pi^2 D_1^3 D_2^3}{3br^3} \tag{5.2}$$

From this simple expression it can be seen that the energy of interaction from which the force derives varies as the 6th power of the diameter and, hence, the diameter alone becomes the critical property together with the saturation magnetization, which should also be maximized.

Fortunately, the material with the highest saturation magnetization is also one of the most economical, i.e., elemental iron. Micron size particles of iron are available commercially from a number of suppliers (e.g., BASF Ag product.[6]), although most of the less expensive forms contain some sort of impurity, often carbon, which reduces the saturation magnetization by approximately 25%. However, fortuitously, the inclusion of impurities such as carbon also improves the corrosion resistance of such materials, hence, enhancing the overall stability of the resulting magnetorheological fluid.

When a magnetic field is applied to a magnetorheological fluid containing micron size particles, the magnetic domains within such particles are readily removed as the particle seeks to align its moment in the field direction. There then exists a very strong force of attraction between the particles, which as indicated above, has to some extent been moderated by the inclusion of thixotropic agents or filler particles. However, once the field is removed, the magnetostatic energy of such particles is such that the domain structure is immediately restored and, hence, no permanent force of attraction between the particles exists. Of course the balance of the domain structure is never perfect and there can be some residual remanent magnetization in the system. However, in properly formulated materials with the correct particle size and dispersion characteristics, this remanent value is minimized.

Magnetic Behavior

Figure 5.3 shows the magnetic behavior of such a magnetorheological fluid. In this case, the material contains 40% volume fraction of elemental iron particles dispersed in a hydraulic oil. As can be seen from the figure, this material consists of 5 μm particles and appears to saturate in a field of about 4 kOe.

We have examined the magnetic behavior of many such fluids and, while not shown in the figure, we find that the magnetization is almost completely reversible for the smaller particles, but for larger particles some residual remanence is almost inevitable. However, we have also found that under modest shear, such residual magnetization is removed almost immediately due to the disorienting effect of the shear forces.

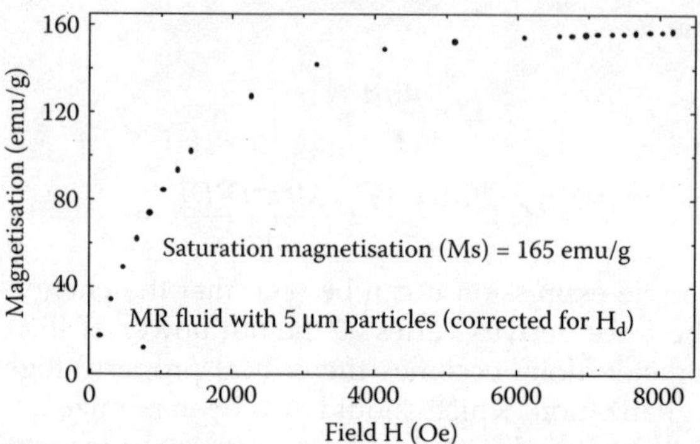

FIGURE 5.3
The magnetization curve of a typical magnetorheological fluid containing 40 vol% of carbonyl iron particles.

The form of the magnetization curve that is obtained depends critically on a number of parameters: the first of these is the particle size. The energy of a magnetic dipole in an applied field is given by Equation 5.2. Hence, given that the magnetic moment depends upon the volume, the energy is greatest for a particle of larger volume and, hence, the magnetization would be expected to saturate more quickly for larger particles than smaller. Experimentally, this is found to be the case. Obviously, the particle concentration not only affects the value of the saturation magnetization and hence the maximum force that can be achieved, but also affects the form of the magnetization curve itself. This is because the particles, once magnetized, interact strongly together. The form of this interaction in such a many-body system is complex and under certain circumstances can be both magnetizing and demagnetizing. However, in general, the particle concentration lowers the initial susceptibility, which after the application of a certain critical field, becomes very large and the material saturates more readily. Obviously, the magnetic interaction between the particles is affected, not only by the particle volume, but also by the interparticle separation, which is itself controlled by the concentration. Similarly, the degree of dispersion in the colloid also affects the form of the magnetization curve as particles that are not separated and dispersed essentially behave as larger particles giving rise to effects described above. Thus, the magnetic behavior of such colloids is quite complex, although the basic principles of this behavior are relatively simple.

Rheological Behavior

Figure 5.4a shows the variation of the viscosity of the same colloid for which the magnetization curve is shown in Figure 5.3 as a function of the shear rate. This graph shows that the material exhibits normal thixotropic behavior as

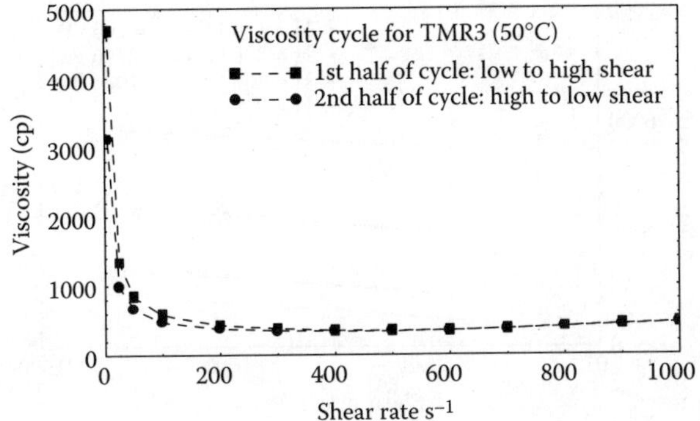

FIGURE 5.4a
Viscosity versus shear rate for a typical magnetorheological fluid in zero-applied field.

expected with a slight hysteresis at low shear, again as expected. Figure 5.4b shows data for the same material under different values of applied magnetic field. Again, at low shear the thixotropic behavior of the fluid is significantly enhanced by the presence of the magnetic field. For example, the low shear viscosity of the fluid in a field of 1.2 kOe is believed to be as high as 10^5 cP, whereas the zero shear viscosity in zero field is only of the order of 5000 cP. Of course, these values have to be extrapolated due to the fact that viscosity cannot be measured in the absence of shear. Under higher shear conditions, it is clear that the effect of the magnetic field is less dramatic than it is at low shear, but nonetheless for a shear rate of 1000 s⁻¹ the change in the viscosity between zero field and 1.2 kOe is of the order of a factor 5.

Of course, for device application, the variation of shear stress with shear rate must also be examined. Ideally, one would wish for an almost linear relationship, which cannot be achieved in a thixotropic fluid. The data in Figure 5.5 shows the variation of shear stress with shear rate for the same

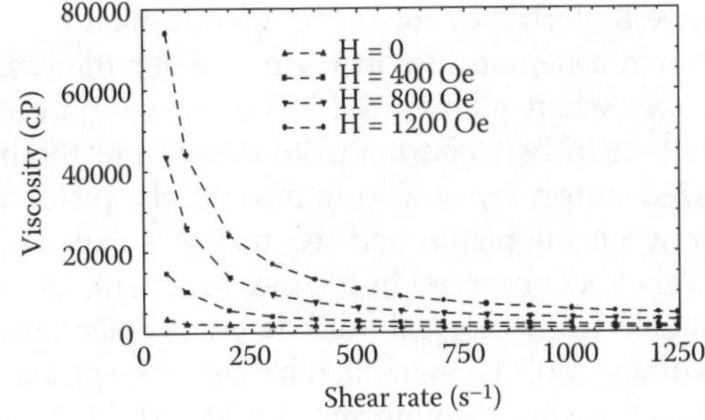

FIGURE 5.4b
The effect of applied DC magnetic field on the viscosity of a typical magnetorheological fluid.

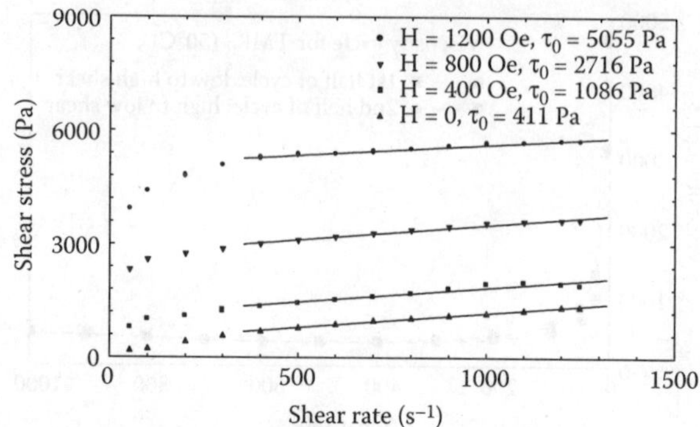

FIGURE 5.5

The variation of shear stress with shear rate for a typical magnetorheological fluid in a range of applied DC magnetic fields.

fluid as examined in the previous section and shows that an extended quasi-linear region does occur for shear rates between a few hundred s^{-1} up to around 1300 s^{-1}, which is the limit of our measurement capability. Similar behavior is observed for all values of field although the extent of the linear region tends to decrease slightly as larger magnetic fields are applied. Given that the magnetic fields are maintaining a strong interaction between the colloidal particles, this type of behavior is to be expected. However, the data shown in Figure 5.5 does indicate that a device with predictable properties could be made from such a fluid.

Electrorheological and Magnetorheological Devices

Early designs of devices that used electrorheological and magnetorheological fluids were generally very similar. Considering the simplest case of a small piston damper, the cylinder of the damper was filled with the appropriate fluid and a field, be it electric or magnetic, was applied in the region of the piston. This is shown schematically in Figure 5.6 for the case of an magnetorheological damper where a small coil has been wound onto the piston to produce a magnetic field between the piston itself and the mild steel body. The coil is energized simply by powering through the piston rod and, hence, the fluid lying between the piston and the casing is expected to exhibit the magnetorheological effect observed in the measurements on the bulk fluid.

The design of an electrorheological fluid device is essentially similar except that a very high voltage must be generated between the piston and the casing, again supplied in some way through the piston rod. In the case of an electrorheological fluid, it is not possible to generate an electric field through the bulk of the liquid and, hence, only that small portion of the liquid lying between the piston and the casing can be activated and take part in the

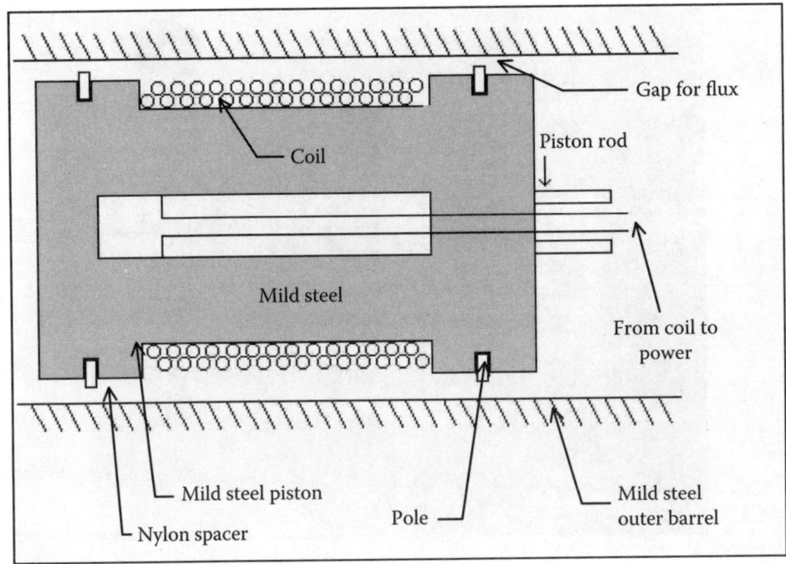

FIGURE 5.6
Schematic design of a traditional magnetorheological damper.

electrorheological effect. However, for the case of magnetorheological fluids, this does not apply as it is relatively easy to generate a magnetic field in a significantly larger volume, particularly when a highly permeable material such as an magnetorheological fluid is present within that volume.

Hence, a revised design of an magnetorheological damper has been developed by Liquids Research Ltd. using computer aided design (CAD) techniques to apply a magnetic field to the bulk of the fluid lying within the piston cylinder. The fluid is then activated directly by the fixed coil and no power is supplied to the piston itself. A photograph of the device, whose performance is discussed subsequently, is at Figure 5.7.

Evaluation of a Magnetorheological Damper

A full evaluation of this device has been undertaken using the facilities of Prodrive Ltd., who have extensive facilities for the testing of dampers. The Prodrive test rig is fitted with sensors to measure displacement and temperature. A power supply delivering a maximum voltage of 14 volts, with a current of up to 4 amps, is available to energize the coil. As will be obvious, these values are almost ideal for automotive application as opposed to the use of very high voltages which are necessary for electrorheological dampers.

The test rig is capable of delivering piston velocities of up to 3 m/s over a 200 mm stroke and can measure damper forces of up to 20 kN. The test

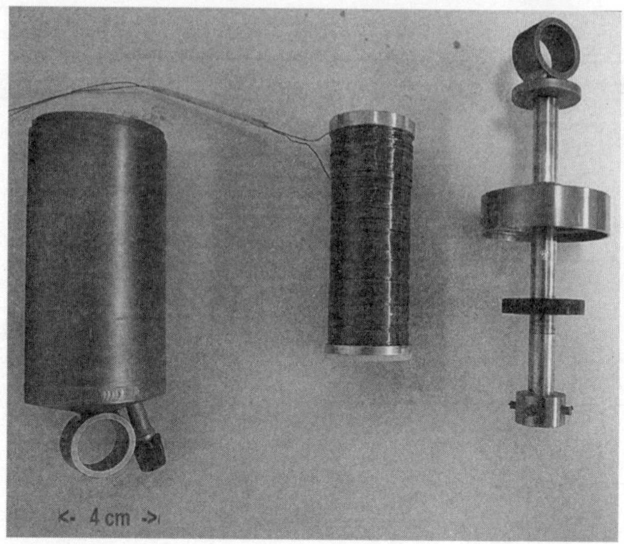

FIGURE 5.7
A novel prototype magnetorheological damper with the field applied to the body of
magnetorheological fluid.

strategy was to determine the damper dynamics and, hence, to determine
the optimum current for the operation of the damper, as well as measuring
the response rate of the damper after the application of a magnetic field, and
determining at least partially the temperature capability of the device.

Damper Test Results—Load versus Velocity

Figures 5.8 and 5.9 show the variation of the load supported by the piston
as a function of the velocity of its movement for a range of currents through
the coil. In all cases, the stroke of the piston is 100 mm. The data in Figure
5.8 shows that the magnetorheological effect saturates for a current of only
1 A through the coil. The higher resolution data shown in Figure 5.9 shows
that, in fact, the vast majority of the magnetorheological effect saturates at
currents significantly less than 1 A, meaning that the device, which is of
similar dimensions to a heavy vehicle seat damper, consumes power at a
rate of less than 10 W. However, the data also shows that there is a significant
zero offset as the velocity of the piston is varied.

The response time of the damper has also been examined using similar
data to that shown above, but monitoring the effect after the current is first
switched on. There is not expected to be any significant rise time in the
magnetic field itself and, hence, any delay in response is due solely to the
effect of the particles aligning with the magnetic field. A summary of the data
obtained appears in Table 5.1 and is displayed graphically in Figure 5.10. The

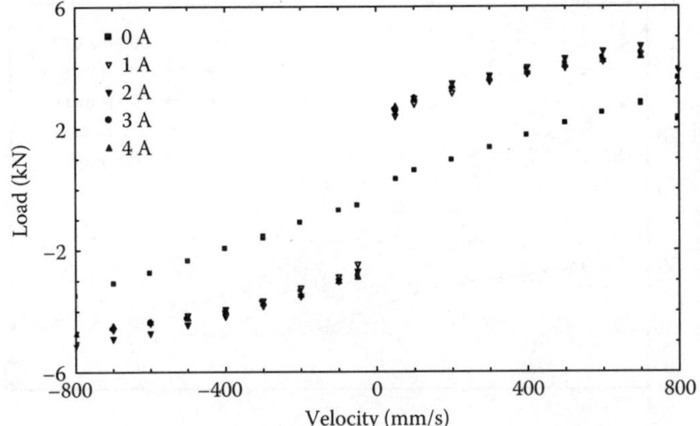

FIGURE 5.8
Variation of load with piston velocity for the damper shown in Figure 5.7 for a range of currents in the coil.

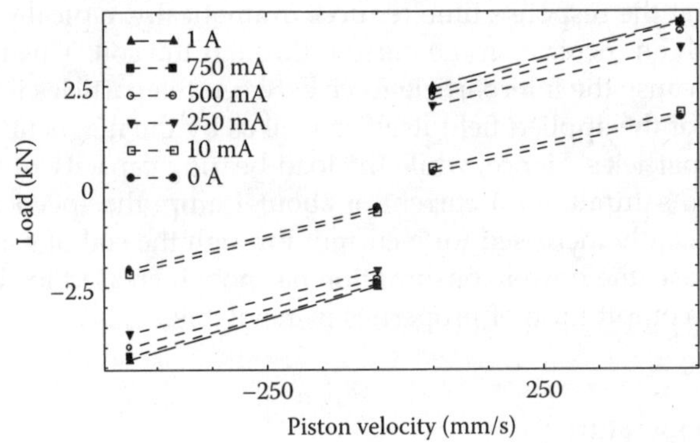

FIGURE 5.9
An expanded view of the data in Figure 5.8 showing the low-field behavior.

TABLE 5.1

Response Time of the Damper Shown in Figure 5.7 for a Range of Applied DC Fields and Piston Velocities

Piston Velocity (mm/s)	1A	2A	3A	4A
100	76 ms	62 ms	44 ms	34 ms
200	76 ms	60 ms	44 ms	40 ms
300	50 ms	46 ms	34 ms	38 ms
400	76 ms	56 ms	44 ms	40 ms
500	86 ms	62 ms	38 ms	42 ms

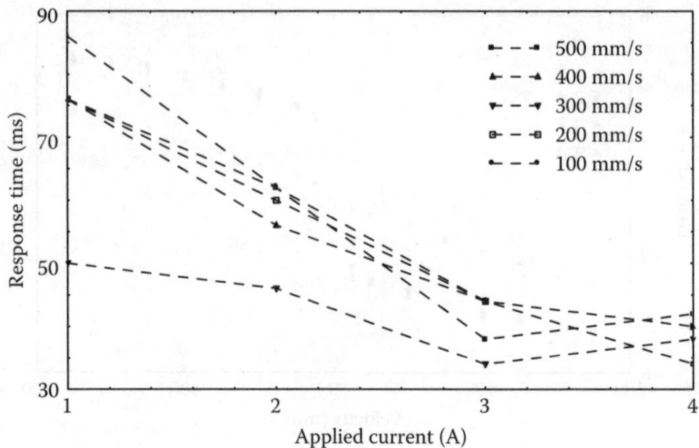

FIGURE 5.10

The variation of response time for the damper shown in Figure 5.7 with current through the coil for a range of piston velocities.

data show that the response time reduces dramatically, typically by a factor of 2 or 3, with an increase in the current through the coil. This is not really surprising because the interparticle force between the particles is affected by the presence of the applied field itself as well as by the magnetic moment of neighboring particles. Hence, while the load-bearing capacity of the damper is observed to saturate for a current of about 1 amp, the speed of response can be significantly increased for a current through the coil of 4 amps. However, in this case, the power consumption has now increased to almost 50 W. Hence, some compromise of properties is inevitable.

Effects of Temperature

The operation of the damper device has been evaluated at temperatures up to 80°C. These data are shown in Table 5.2. In this test, the damper is cycled in simple harmonic motion and the temperature stepped in 10°C intervals. Temperature causes the peak value of the load to drop as the viscosity is

TABLE 5.2

Damper Performance as a Function of Temperature

Temperature (°C)	Load kN (Extension)	Load kN (Compression)
30	4.507	−6.310
40	3.470	−5.228
50	3.020	−4.777
60	1.037	−4.417
70	0.721	−4.191
80	0.586	−4.011

lowered. However, the nature of the simple harmonic motion is maintained, and the behavior is reproducible and predictable. Of course, the viscosity of any material changes with temperature and the thixotropic effects in magnetorheological fluids give rise to a greater temperature coefficient of viscosity than is the case in normal hydraulic fluids.

An extensive range of other tests has also been done to examine the exact conditions necessary to control this small prototype damper. Using the sophisticated control of the test rig removes the zero velocity effect at a peak velocity of 800 mm/s. Predicting the behavior of the damper, also removes other effects such as bump and rebound, and the damper rate being proportional to the velocity meant that the load gradients of the damper itself can be controlled. This is particularly true at low operating speeds.

Other Devices

As indicated in the introduction, a number of other possible devices can be envisaged. Of particular importance for automotive applications are slip clutches, which can have significant energy consumption implications for the control of, for example, the cooling fan on large vehicles and also in the front to rear differential on 4-wheel drive vehicles. The principle of the design of a magnetorheological slip clutch is shown in Figure 5.11. A fixed pole piece with a coil is located in one side of a barrel with a steel flywheel positioned within the same cavity, which is subsequently filled with magnetorheological fluid. Please note that this diagram is schematic and that in practice, the steel flywheel and the pole piece and coil would be physically

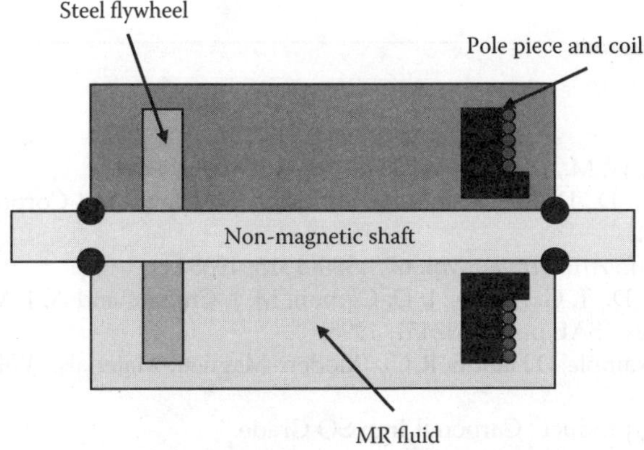

FIGURE 5.11
Schematic design of an Magnetorheological clutch/brake.

much closer together. It should also be noted that such a device is already available commercially from the Lord Corporation of Carey, North Carolina, and has found application in human exercise machines when operating as a brake.

When the coil is activated, the magnetorheological effect couples the flywheel to the fixed pole piece, thereby providing variable and extremely well controlled coupling. Of course, such devices are subject to large centrifugal effects at high speed and special design considerations must be applied in order to limit these effects, which can destabilize the magnetorheological fluid. One example of a possible solution to this problem is the design of labyrinthine plates within the slip clutch arrangements.

Summary

This chapter has described briefly, the historical development of magnetorheological and electrorheological fluids, which are now commercially available. The similarities and differences between these two classes of similar material have been highlighted and the properties of recently developed magnetorheological fluids discussed. The fundamental mechanisms that provide the magnetorheological effect and can lead to the production of materials with long term stability have been described together with both the magnetic and rheological properties of such materials. A detailed evaluation of a prototype magnetorheological damper, such as that used in the seat mountings of heavy vehicles, has been described together with a brief description of other possible devices for use in the automotive sector and elsewhere, such as slip clutches.

References

1. Winslow, W. M., *J. Appl. Phys.*, Vol. 20, 1137–1240, 1949.
2. Carlson, J. D., D. M. Catanzarite, and K. A. St. Clair, Lord Corporation, Cary, NC 27511.
3. Rabinov, J., *AIEE Trans.*, Vol. 67, 1308–1315, 1948.
4. Weiss, K. D., T. G. Duclos, J. D. Carbon, M. J. Chrzan, and A. J. Margida, Soc. Auto Engs., SAE paper 932451, 1998.
5. See, for example, O'Handly, R. C., "Modern Magnetic Materials," Wiley, Hoboken, NJ, 2000.
6. BASF Ag product "Carbonyl Iron SQ Grade."

6

Impact Loading

Nik Petrinic

CONTENTS

Introduction

During the past decade numerical crashworthiness simulation for vehicle and occupant safety protection has been greatly integrated into the vehicle design process, and is now a mature technology. Manufacturers have considerably reduced costly and time-consuming prototype crash-testing programs, and the physical testing is needed only to validate the design based upon computer simulations. This has led to a reduction in development time (from 5 to 3 years on average) and a substantial cost saving (over 30%). In addition, the simulation-based design has led to consistently improved vehicle designs and a reduction in road deaths of over 30%.

Achieving reduction in fuel consumption and CO_2 emissions while improving vehicle performance and safety is a major challenge facing the automotive industry today. Improvements in aerodynamics, fuel, motor, and drive chain technology have provided an over 20% reduction in emissions over the past ten years. However, these potential benefits have been negated by a 20% increase in average vehicle weight during the same period.

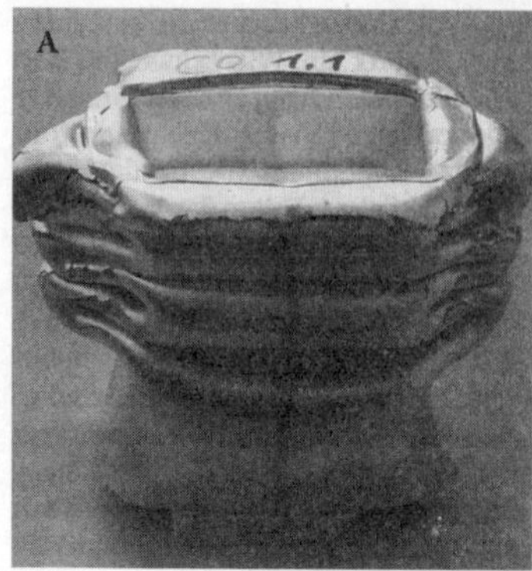

FIGURE 6.1

Examples of automotive components from high-strength, lightweight materials: (a) an axially crushed aluminum tube showing extensive metal tearing, (b) a magnesium door, and (c) a high-strength steel rear axle.

The European Association of Automobile Manufacturers has a stated commitment to limit average output of CO_2 to 140 g/km per vehicle by the year 2008 (currently 190 g/km).[1] If this objective is to be achieved, improvements in all technological areas and, in particular, significant weight reductions, will be essential. As a result, the industry is increasingly using high-strength metals (high-strength steels, aluminum, and magnesium [Figure 6.1]) and low-weight polymers in order to achieve required weight reductions by compromising neither performance nor occupant/pedestrian safety. It is realistic to expect low-weight materials to reduce average vehicle weight by 15%–20%, which in turn will reduce fuel consumption by 0.5–1 Ltr/100 km. This is estimated to reduce considerably the total CO_2 emissions (by 30%) over the lifetime of the car.

The present commercially available software for numerical simulation of car crash events have proven remarkably reliable for designs based upon conventional ductile steels in which the principle response to impact loading comprises large inelastic material deformation and structural bending, buckling, and collapse of sectional members.[2] However, the algorithms employed in such software are struggling to predict the dynamic material failure (including crack propagation) observed in experiments with newly utilized high-strength steels or lightweight metallic and nonmetallic materials for automotive applications.[3–6] These materials respond to dynamically applied loading by illustrating considerably lower ductility when compared to conventional automotive steels, thus changing the principal structural response to impact loading. Similar difficulties are encountered in predicting failure in jointing systems, which also play a significant role in crashworthy designs.[7] Consequently, present simulation results are not adequately predictive. As a result, in order to avoid a return to costly prototype-based designs, it is imperative that failure modelling capabilities are improved if effective utilization of advanced materials is to be continued without compromising safety.

Adopted Methodology

In order to enable progress toward improved predictive modelling of deformation and failure of lightweight materials subjected to impact loading, the following integrated experimental/numerical methodology has been adopted by researchers in industry and academia interested in automotive crashworthiness.

1. New experimental programs for selected advanced materials must be devised in order to provide required information on the dependency of material behavior to process history, temperature, and rate of loading. Such programs must encompass intelligent-selective testing rather than mass testing, and must provide manageable procedures and necessary data that enable determination of required constitutive parameters. New methodologies must be established to characterize the deformation and failure of jointing systems.

2. New constitutive models that can include process history, temperature effects, and can adequately describe the evolution of material state at required scale during loading, must be developed and implemented into existing software. Appropriate criteria for crack initiation and crack propagation (transition from continuum to discontinuum) must be incorporated into constitutive models and novel techniques, just as automatic remeshing around evolving discontinuities must be used for improved accuracy. Extensive validation against experimental data should be carried out during development and implementation in order to ensure that mathematical models adequately represent the observed physical behavior.

3. Inverse modelling techniques must be employed to fully quantify the response of different materials to impact loading with respect to newly developed numerical algorithms, as not all constitutive parameters will be directly measurable in experiments.

The development framework representing the basis of such design methodology can be described as an iterative process, part of which can be greatly automated by employing inverse modelling techniques based upon optimization and stochastic analysis. An example of such development framework for research into behavior of automotive materials subjected to impact loading is illustrated in Figure 6.2. In this approach, the investigation starts with small-scale laboratory experiments in order to provide both qualitative and quantitative information on the observed behavior. The next step comprises the theoretical (mathematical) abstraction of the selected phenomena and is followed by development and implementation of numerical algorithms into chosen computational modelling framework. Direct comparison between the results of experimental measurements and numerical simulations can be used to determine non-measurable modelling parameters, and can also provide information on the accuracy of newly developed numerical modelling tools.

Moreover, the inverse modelling for identification of nonmeasurable modelling parameters can establish the adequacy of newly developed algorithms for solving given problems. This follows from the ability of optimization and stochastic methods to separate regions within analyzed parametric spaces if the employed models cannot adequately represent the observed behavior.

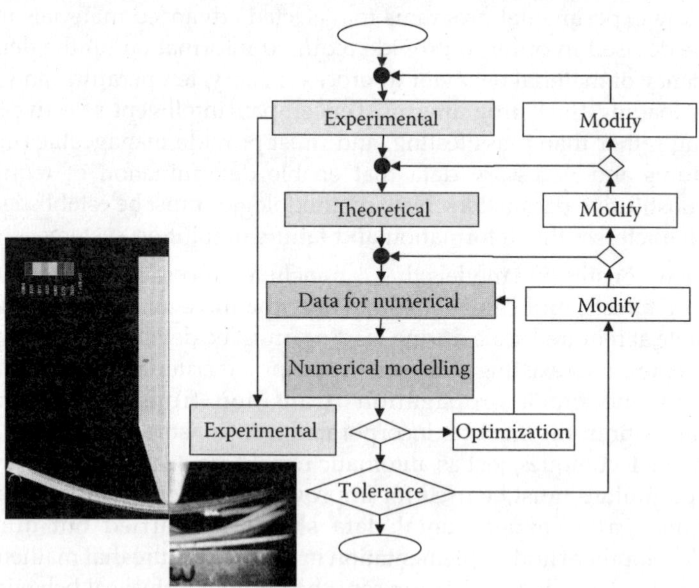

FIGURE 6.2
Integrated experimental/numerical approach to predictive modelling of deformation and failure of automotive materials.

Research Focus

Experimental Focus

In order to describe the behavior of automotive materials subjected to impact loading, a set of loading rigs should be employed that can generate material states similar to those in components of structures involved in impact. This mainly means enabling the application of loading at strain rates between 10^{-4} s^{-1} to 1500 s^{-1}. These are the typical limits observed in instrumented car crash events and their numerical simulations. In order to replicate the observed phenomena in laboratory conditions, the classical quasi-static (screw-driven) and dynamic (servo-hydraulic) loading rigs should be complemented by a split-Hopkinson-bar (SHB) apparatus (Figure 6.3) that can deliver loading at strain rates of up to 10^4 s^{-1}. In the case of automotive materials, the development of experimental techniques capable of delivering loading at high rates of strain involves working with specimens obtained from rolled materials and, as such, comprises numerous challenging tasks from gripping the specimens to interpreting the results of measurements. The use of all three types of loading rigs enables loading at three distinct rates of strain, which in the case of a typical automotive steel alloy, results in a behavior illustrated in Figure 6.4. Significantly different behavior is observed when aluminum alloy is tested at the same conditions (Figure 6.5). Aluminum does not show any relevant rate dependency at levels of strain below the onset of material instability (necking). The rate dependency is observed in terms of an increase in strain at specimen fracture.

Numerical Focus

The constitutive modelling of deformation and failure of materials subjected to impact loading will be as accurate as the discretization models allow. This is due to the inability of constitutive models to provide adequate predictions

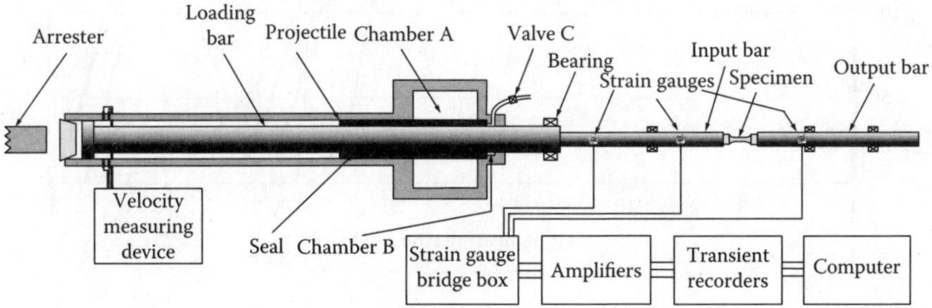

FIGURE 6.3
Tensile split-Hopkinson-bar apparatus.

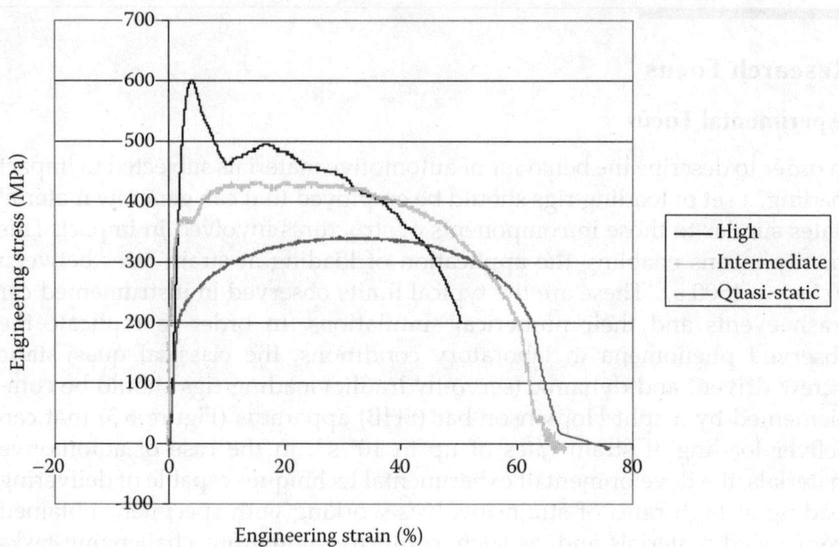

FIGURE 6.4
Rate dependency observed in typical automotive steel alloy.

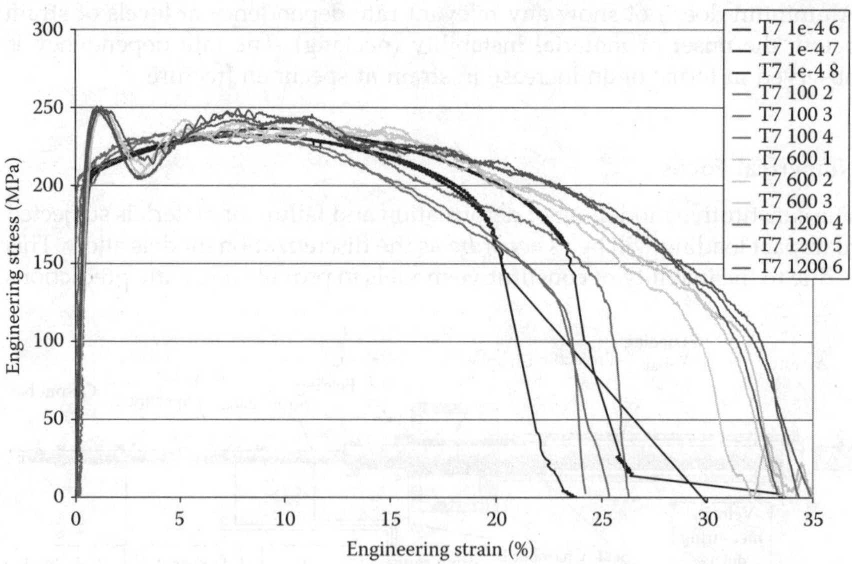

FIGURE 6.5
Rate dependency observed in typical automotive aluminum alloy.

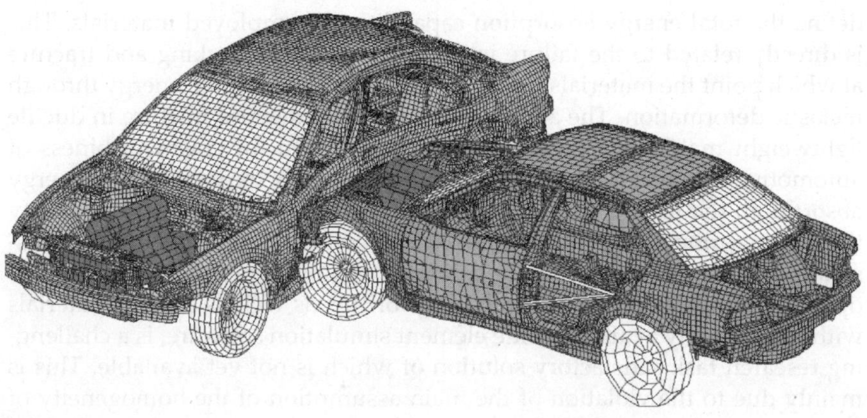

FIGURE 6.6
An example of large-scale crash simulation.

if forced to operate at representative volume element sizes considerably larger than the scale at which the processes characterizing the observed physical phenomena happen. For example, the required size of representative volume element needed to adequately simulate linear elastic and moderate strain inelastic behavior of steel automotive components, may be of the order of 1–5 mm. On the other hand, if the modelling of localization and failure is attempted in the same material, the size of representative volume element should be of the order of 0.05–0.1 mm. This poses high demands on the manufacturers to provide extreme computing power capable of large-scale crash simulations, as illustrated in Figure 6.6.

$$\dot{\varepsilon} = 1500\,\mathrm{s}^{-1}$$

$$\dot{\varepsilon} = 100\,\mathrm{s}^{-1}$$

$$\dot{\varepsilon} = 10^{-4}\,\mathrm{s}^{-1}$$

In addition, the large scale crash models will not predict accurately the material response unless the process history is taken into account in simulations. This means that a set of complex modelling operations is needed that will enable mapping of the existing results following forming (stamping) simulations onto crash models. This way, the material states in structural components are taken into account, thus reflecting the variability in material responses depending on the process history.

The constitutive parameters for use in crash simulations can be obtained from experiments on specimens obtained from virgin and processed material samples. Alternatively, the constitutive models used in simulations of forming processes should provide all required information needed by the constitutive models used in crash simulations.

The most critical aspect of numerical simulations of the behavior of lightweight automotive materials subjected to impact loading is the ability to define the total energy absorption capabilities of employed materials. This is directly related to the failure in materials through cracking and fracture at which point the materials lose their capability to dissipate energy through inelastic deformation. The ability to predict cracking and fracture in ductile lightweight materials represents the key to design for crashworthiness of automotive structures. In order to estimate accurately the total level of energy absorption, the failure criteria (discrete and continuous) must fully satisfy the energy principles of thermodynamics in which the formulations of dissipation potentials must be fully validated against experiments. The development of failure criteria for both quasi-brittle, as well as ductile materials within the widely available finite element simulation software, is a challenging research task satisfactory solution of which is not yet available. This is mainly due to the violation of the main assumption of the homogeneity of deformation within the representative volume element employed by the continuum mechanics and finite element method.

Summary

The behavior of selected automotive materials subjected to high-rate loading has been described in the context of computer-aided design for automotive crashworthiness. The capabilities of currently used laboratory equipment and numerical simulation tools have been outlined in order to provide an overview of the design methodology presently employed. The focus is placed on evaluating the response of the materials rather than the structure to impact loading in an attempt to quantify the capacity to dissipate kinetic energy during crash events.

At times when environmental issues play a dominant role in engineering design, the consideration of lightweight materials represents one of the logical steps toward low fuel consumptions and reduced CO_2 emissions. However, the introduction of lightweight materials requires rethinking of the currently employed design procedures, as the simple substitution of new materials into traditional structures does not automatically bring satisfactory results. Such a radical step is only possible if adequate experimental and computational supports are available, both of which also have to be modified and adapted during the redesign process.

References

1. CO_2PERATE: The R&D programme on Automotive CO_2 Emissions reduction, Programme manager: Dr Ulf Palmer, 1999.
2. Scharnhorst, T., and Schettler-Köhler, R., "FEM-CRASH Experiences at Volkswagen Research," *Computational Mechanics*, 75–86, 1986.

3. Lemaitre, J., and Chaboche, J. L., *Mechanics of Solid Materials*, Cambridge University Press, Cambridge, UK, 1994.
4. Gurson, A. L., "Continuum theory of ductile rupture by void nucleation and growth, Part 1: Yield criteria and flow rules for porous media," *J. Engng. Mates. Tech., ASME*, Vol. 99, 1977.
5. *PAM-CRASH™ FE Code*, Engineering Systems International, 20 Rue Saarinen, Silic 270, 94578 Rungis-Cedex, France.
6. (*Strain rate testing of steels: German Automotive FAT group*), Fraunhofer Institut fuer angewandte Materialforschung (IFAM), Ermittlung dehnungsgeschwindigkeitsabhaengiger Werkstoffkenngroessenan Karosserieblechen (Abschlussbericht zu FAT-Auftrag), 1993.
7. (*Spotweld testing: German Automotive FAT group*), Oeter, M., Aufbau und Inbetriebnahme eines Hochgeschwindigkeits-material-und Bauteilpruefsystems, LWF Paderborn, 1999.

7

High-Temperature Electronic Materials

Colin Johnston

CONTENTS

Introduction

The modern automobile relies heavily on advanced electronic systems for vehicle performance and control, performing functions such as fuel injection and emission control, anti-skid braking, active suspension, and electronic transmission control (see Figure 7.1). A recent study by the Freedonia group estimates that the growth for automotive electronics alone will increase from $1,208 per vehicle in 1999 to $1,864 per vehicle in 2009.[1] While some of this increase is due to the evolution of telematic systems, most of this growth will be due to hybrid vehicle electronics, collision avoidance and protection systems, electronic steering and vehicle stability, and powertrain management with the incorporation of new systems like drive-by-wire control systems (throttle, steer, brake, shift, and suspension by wire), collision avoidance systems (automatic braking, steering, and throttling with radar), and advanced energy systems (42 volt, fuel cell controllers, and advanced energy converters).[2] This accelerating trend toward more advanced electronics

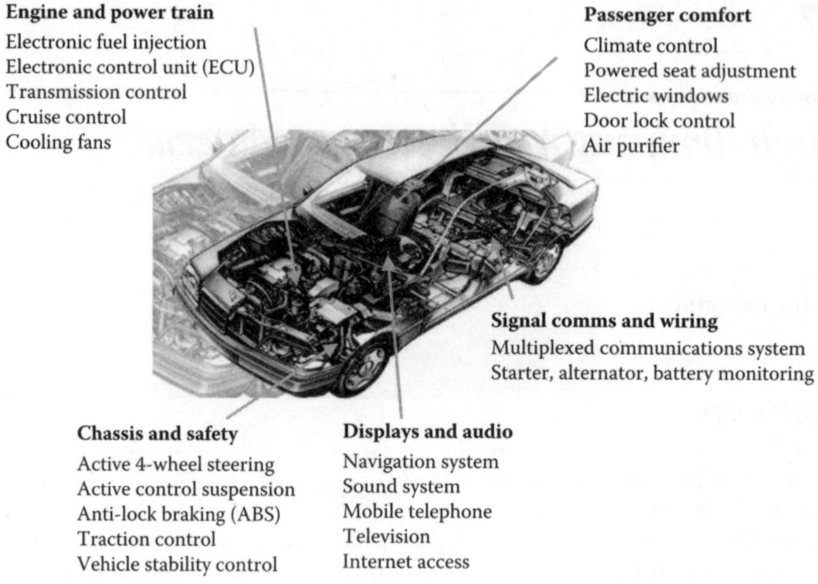

FIGURE 7.1
Electronic systems in a modern automobile.

will increase the use of vehicle electronics systems to an unprecedented level.[3]

This unprecedented technological growth in automotive electronics is best illustrated by the evolution of one particular subsystem—powertrain management. Engine and transmission management controllers now support increased feature content, with additional "smart" subsystems added to provide detailed and fast electromechanical interfaces.[4–6] Modules like voltage regulators, airflow meters, power switching, and smart solenoid switching systems help the electronic control systems to monitor powertrain performance and adjust mechanical operations.

This trend to smart sensing, processing, switching, and driving provides vehicle suppliers with two strategies for the overall systems electronics. One option is to develop a very complex powertrain controller module capable of monitoring and adjusting a large number of inputs and outputs in real time. This option is becoming increasingly difficult as systems require more feature content, which in turn, increases module packaging size and increases vehicle wiring.

Another, more attractive option, is to create a number of smart powertrain modules each performing a series of specific operations. Many companies are now moving to these types of systems (called distributed controls), which use a series of "mechatronic modules" and smart actuators in conjunction with a smaller central controller unit. Such systems are integrated through a communications bus (e.g., CAN), giving more direct operational capability for the controller system. Besides reducing the overhead of the local control

TABLE 7.1

Typical Temperature Ranges Encountered by Mechatronics in a Modern Automobile

On engine	150 to 200°C
In transmission	150 to 200°C
On wheel—e.g., ABS sensors	150 to 250°C
Cylinder pressure sensing	200 to 300°C
Exhaust gas sensing	up to 850°C with ambients up to 300°C

algorithms in the PCM, this architecture also reduces the associated complex wire harnesses and connectors running over the vehicle. Reducing wire harnesses and connectors decreases cost, weight and EMI problems, signal attenuation, and delay, which also decreases the reliability risk associated with wiring harnesses and connectors.[7-12] Unfortunately, many of these mechatronic modules are located in hot locations requiring high temperature electronics (HTE) such as on-engine, on-transmission, and perhaps soon, on-wheels for brake-by-wire and steer-by-wire. Table 7.1 provides a general range for temperatures in potential mechatronic locations. In addition to temperature, automotive mechatronics must also exhibit high vibration and shock tolerance, and high reliability.

Market Perspective

The automotive sector represents the largest potential single market opportunity for high temperature electronics.[13] High temperature electronics components could significantly improve system design and enhance fuel efficiency leading to overall higher performance vehicles. However, the automotive sector represents the most stringent pricing pressure of any market likely to be covered by high temperature electronics. In addition, demands on the reliability of the high temperature electronics component should not be ignored—most car manufacturers demand components that last the lifetime of the vehicle, which is not insignificant.

Nevertheless, since the total available automotive electronics market is so large (ca. > $14 billion/year), even a small fraction of this represents a large market opportunity (currently ca. $120m rising to ca. $550m by 2008) to high temperature electronics providers, as shown in Figure 7.2. However, much development needs to be undertaken not only in basic component architectures, but also on mass production technologies for high temperature electronics before any significant penetration can be achieved.

Advanced automobiles are already heavily reliant on electronics and sensors. These include sensors for engine speeds, angular position, ABS, exhaust gas, power steering, engine condition monitoring, and electric windows. High temperature electronics to be co-located with the sensors is currently being

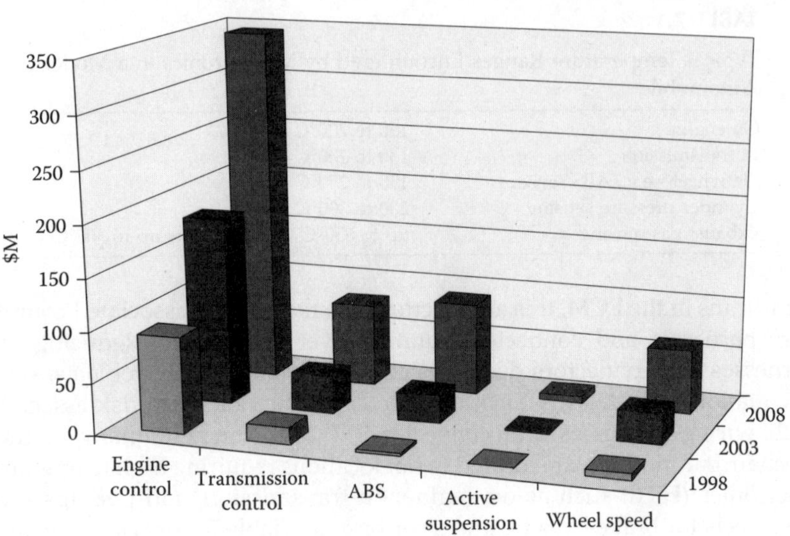

FIGURE 7.2
The automotive market for high temperature electronics.

developed. Furthermore, high temperature mechatronic systems will become more common, offering improved safety and comfort, and reduced costs.

The major market driver for the development of high temperature electronics for the automotive market sector is legislation to reduce emission and improve engine efficiency and, hence, economy. Another driver is to reduce manufacturing costs for the automotive manufacturers (see Figure 7.3). As a consequence of increased utilization of high temperature electronics in automobiles,

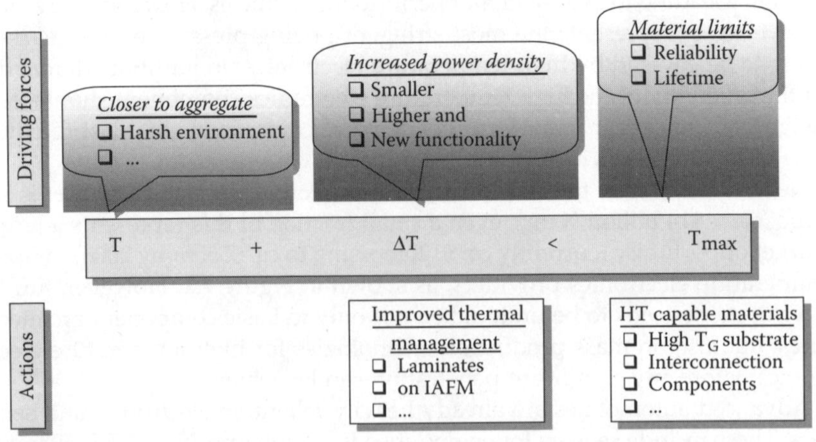

FIGURE 7.3
Rationale for high temperature electronics development in the automotive sector.

there will be consumer benefits with improved reliability and increased efficiency that should work to reduce the average cost of ownership in terms of both reduced running costs and reduced maintenance requirements.

Module Technology

Currently, most harsh environment electronics are designed to withstand a temperature range of –40°C to +125°C. These systems must also typically meet automotive vibration requirements while exceeding 10 years and 100,000 miles of operation. To limit the effects of the vehicle environment, electronics modules are often separated from the mechanical systems, which they control. Locations like vehicle firewalls (Figure 7.4) and fender wells offer the ability to sink module-generated heat while reducing exposure to temperatures created by the mechanical systems, and allowing some access to airflow available under-the-hood.

Unfortunately for designers, future vehicle electronic modules will be physically integrated with the mechanical systems they are intended to control. This will eliminate many of the module separation opportunities, and will place the electronics directly in the thermal generation areas of the mechanical systems and subject them to increased temperatures (Figure 7.5).

The next generation of automotive electronic control units (ECU) is expected to be mounted close, or directly onto, the actuator—this means, for example, directly at the engine, into transmission, or near the brake disk. This localization of electronic control units represents the evolutionary step from the distributed mechanical system toward a functional integration of mechanics and electric with electronics.

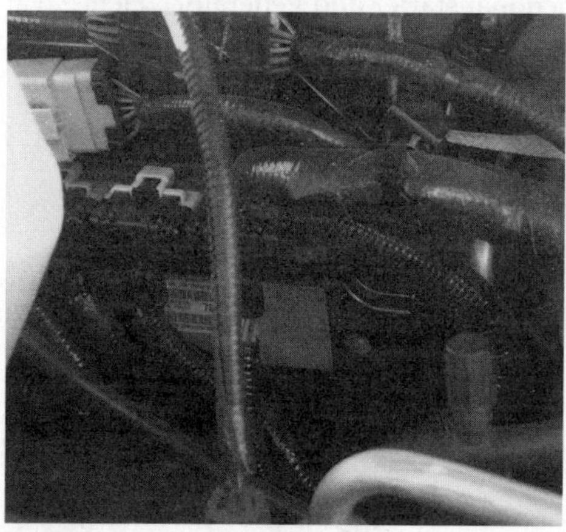

FIGURE 7.4
Electronic control on firewall.

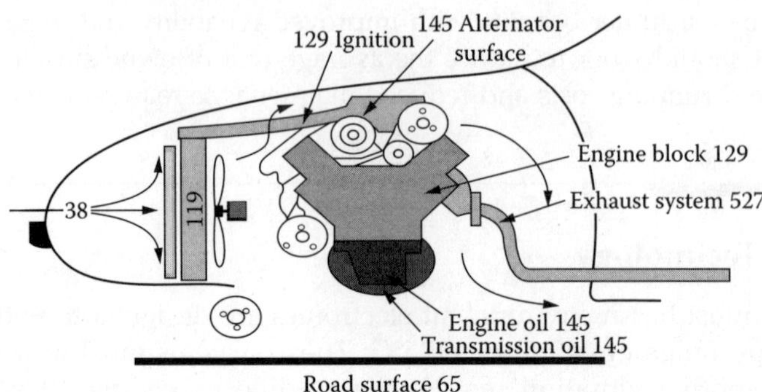

FIGURE 7.5
Typical under-hood temperatures.

For engine controllers, various reasons suggest having the electronic control unit closely mechanically coupled with the engine itself (see Figure 7.6): lack of space in the engine compartment, reduction of wiring length and complexity, reduction of EMI/EMC related problems, improved intake air cooling action. The economical and logistical impact of treating the engine as self-standing component-ready, and fully tested and calibrated, for production at the end of the assembly line are also significant.

With respect to gasoline engines, the control of the intake/exhaust valves, by means of sophisticated electronic systems, will be mandatory for further emissions reductions. Due to the large amount of electrical power to be managed and the request for miniaturization, HT solutions are under evaluation. In order to increase torque, performances and efficiency of thermal engines, the functionalities of starter and generator/alternator (see Figure 7.7) are going to be integrated and realized within one electrical machine (multifunctional starter and generator). Hence, various features will be allowed, such as: start-up directly, start/stop, continuous power generation,

FIGURE 7.6
Engine-mounted electronic control unit for diesel common rail application.

FIGURE 7.7
Multifunctional starter–generator system.

energy recovery, and lowering idle-speed. The power electronic control unit required for the management of the machine will be installed in a position that optimizes losses and physical connections (wire harness).

The Train Control Unit (TCU) also represents an advanced mechatronic system for automatic gearbox control, directly mounted inside the gearbox, able to operate in harsh environments (Figure 7.8).

Silicon Development

Due to the strict low-cost requirements in the automotive industry, only silicon-based solutions seem to be promising for mass production targeting the high-volume, cost-sensitive automotive market. From this point of view,

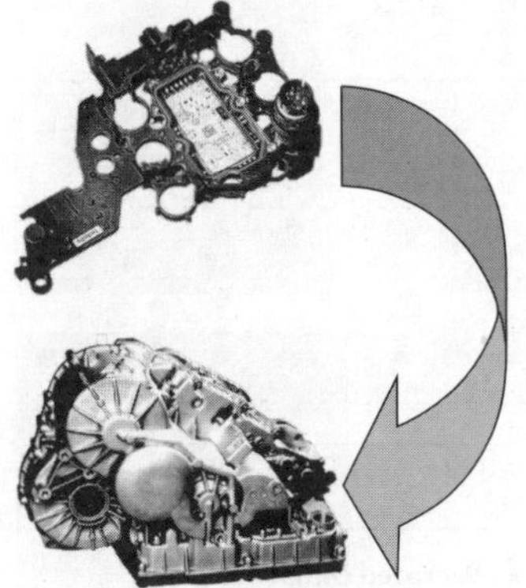

$$T_a < 110°C: 4000 \text{ h}$$
$$110°C < T_a < 125°C: 1000 \text{ h}$$
$$125°C < T_a < 135°C: 100 \times 1.0 \text{ h}$$
$$135°C < T_a < 140°C: \ 30 \times 0.5 \text{ h}$$
$$T_a \text{ up to } 160°C \text{ for next generation}$$

Mechatronic TCU (prodmod)
mounted inside gearbox

FIGURE 7.8
Mechatronic Train Control Unit.

wide-bandgap semiconductors do not represent an affordable solution for the short-medium term. However, with respect to silicon-based technology, silicon-on-insulator (SOI) is rapidly growing and, as a consequence of this rapid growth, the cost of silicon-on-insulator devices is becoming more attractive. Some components in silicon-on-insulator are already available on the market (e.g., CAN controller) because of their robustness and immunity to noise, in addition to the well-known behavior as a function of temperature. To reach the demanding cost target in the automotive, standard designs and silicon CMOS bulk-technologies will be reused as much as possible, by high temperature hardening. As μCs, signal integrated units (logic, memory), ASICs and related driver integrated units are key components for the realization of control units, the following activities will be performed by major silicon makers:

- Adaptation of CMOS technologies for the realization of 16/32 bit μC for $T_j \geq 150°C$
- Process steps tuning and optimization of threshold voltage
- Library optimization for high temperature (design, layout, specific cell)
- Evaluation and realization of test technologies/equipment for HT applications
- High temperature technology qualification
- Evaluation and validation of high temperature bonding metallizations

The evolution of bare die (as well as packaged) components operating at temperatures well above 125°C° is described in Figure 7.9. The development

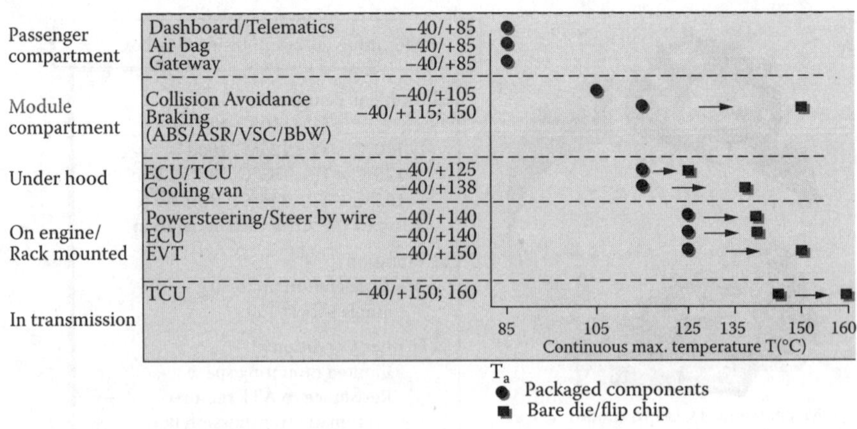

FIGURE 7.9

Evolution of packaged components versus bare die/FC as a function of temperature in the automotive.

TABLE 7.2

Bare Die: Proposal for Temperature Requirement over
Operating Time (Lifetime = 15 years, Operating Time =
400–600 hr)

Temperature (°C)	Operating Time (%)
−40 to 80 °C	10
80 to 110 °C	80
110 to 140 °C	9
140 to 150 °C	1

of electronics and packaging systems with an extended temperature range above 150°C require the adaptation of present packaging and interface technologies. In particular, the development and evaluation of those technologies to fulfil requirements, like die attach, chip size packaging, flip-chip assembly (solder bumping), underfillers, and HT molding (green plastic package) will be needed together with a realistic temperature requirement over operating time (the so-called mission profile, see Table 7.2), as well as acceptable methodologies for testing and reliability assessment.

Packaging and Interconnection Technology

Existing packaging and interconnect technologies do not meet the increased requirements of higher temperatures and harsh environments found in automotive applications. New technologies, i.e., other materials and adapted production processes, must be developed. An exchange of technologies is, however, to be expected, especially in the following areas: wire bonding, substrates, soldering or gluing, and encapsulation. The main issues for packaging materials operating at higher temperatures are:

Adaptation of material combinations with respect to their coefficient of thermal expansion (CTE). This is necessary to reduce mechanical stress caused by thermal mismatch.

Thermal stability of the used materials. This means for metals and their alloys not only the resistance against any kind of oxidation, but also against the formation of critical intermetallic phases. The plastic materials must not be degraded by cracking their chemical composition or by changing their structural arrangement.

With respect to packaging issue in general, there is a clear trend toward a progressive miniaturization obtained through area array devices, due to high density integration. HT applications will be dominated by bare die, but the increased rigidities of the packages, and interconnection will require actions on reliability (e.g., underfilling). LTCC substrates have been mainly used so far in ABS applications, where good thermal resistance combined with the capability to withstand high operating temperatures compared to organic boards, are

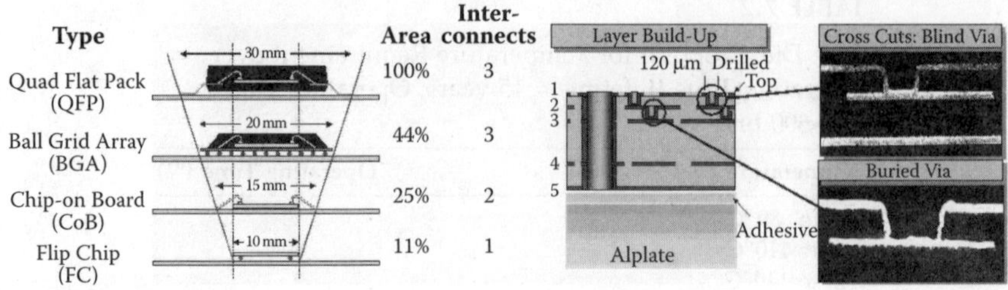

FIGURE 7.10
Packaging evolution and HDI technology for automotive.

required. The thermal conductivity can be further improved using thermal vias in the substrate for direct heat paths. Flip-chip assembly using thermal bumps, in addition to the electrical contacts, will improve the thermal management. Due to the smaller difference in the coefficient of thermal expansion of silicon and LTCC compared to organic materials, a reliability improvement can be expected, as well as the decreased water vapor. The DBC technique is often used in power applications where heat transfer and ohmic electrical characteristics are key parameters (e.g., ignition, injection, electrovalves control).

The advent of high density interconnection technology from telecommunications (see Figure 7.10) based on advanced organic materials, will enable the highest amount of connections on the circuit board in the form of microvias. The characterization and optimization of high density interconnection for automotive will represent a major step toward an innovative technology, thus enabling a substantial miniaturization of the electronics systems: the application temperature is limited by base material (T_g of epoxy resin, solder mask) and adhesion strength of laminates.

By selecting and evaluating new materials, new standards proposed by the EU have to be observed. In the case of printed circuit boards, halogen-free nonflammable materials have to be considered, as well as recycling issues (WEEE directive).

Assembly Process

Robust interconnect technologies are needed to realize electronic systems that are capable to withstand high temperatures. The development and characterization of suited substrates (organic or ceramic) are the basis for assembling electronic control units. In addition, solder materials, as well as soldering techniques, have to be provided that fit well into existing manufacturing processes. It is most likely that these solder materials are lead-free solders. Finally, an adequate housing technology has to be selected and qualified to protect the electronics from the environmental impacts, and to dissipate the heat generated by the components. If active cooling is involved, the housing has to be an efficient interface to the cooler. Thermal simulation is needed to find out if the requirements of the specific application are met by packages, substrates, and interconnect technology. Once the automotive electronic system has been conceived, several types of devices and packages must be assembled onto the same substrate. Hence, interconnect technologies will have to contact:

- Logic devices (high number of inputs and outputs and fine pitch)
- Lower components (high currents, low impedance, back side potential, heat removal)
- Passive components (high mass leads to vibration and causes severe mechanical stress)

Reliability

The required lifetimes for electronic components in automotive applications range from typically 5000 hours for passenger cars to 20,000 hours for commercial vehicles. Specified failure probabilities for electronic control units typically vary between 100 and 500 out of 1 million in 10 years. Some typical results are shown in Table 7.1

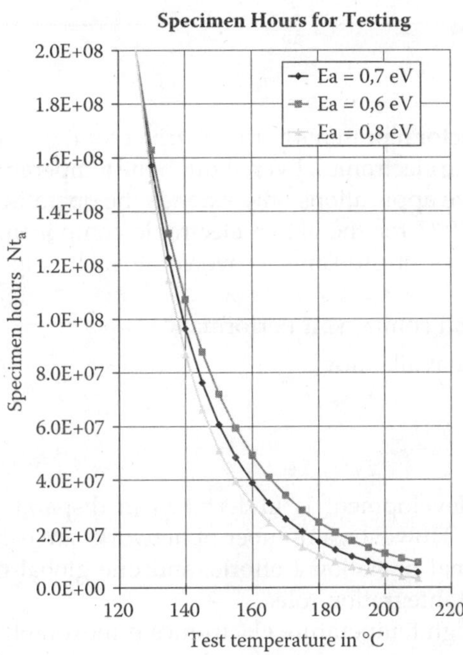

FIGURE 7.11

Requested number of tested specimen hours to obtain a failure rate of 10 fit at 125°C.

In order to ensure a certain level of quality and reliability, defined stress conditions for the qualification testing of devices for automotive applications have to be performed. These accelerated tests comprise electrical, environmental, and mechanical procedures like latch-up, electro-thermal induced gate leakage, temperature cycles, autoclave, temperature-humidity bias, power temperature cycling, solderability, bond shear, vibration, mechanical shocks, etc. Developing related test standards requires the knowledge of the underlying failure mechanisms. Generally, these mechanisms are investigated by accelerated testing and, subsequently, acceleration factors are calculated.

On entering the high temperature regime, further acceleration by simply increasing the temperature becomes a questionable approach. It is still uncertain whether such a methodology can be applied without changing the failure mechanisms. In the future, new applications will run closer to the physical limits of the devices and systems. This means that current qualification tests will no longer be sufficient for some applications. Since there is reduced safety margin for accelerated testing, new procedures for qualification must be developed. Ideally, these new qualification tests should also result in an overall reduction in the qualification effort by, for example, employing simulation.

Summary

The automotive sector represents a critical driver for the wider scale adoption of high temperature electronics. Even though the temperature range encountered in automotive applications only extends the operating environment to a maximum of 150°C for the active electronic components, there is still a considerable number of problems to overcome, including:

- Fundamental component performance
- Component availability
- Packaging
- Reliability testing

Currently, most development is undertaken in disparate groups offering bespoke solutions. However, a number of networks are being established to try to focus regional or national efforts, and one global network, HITEN, performs a critical integration role.

The future for high temperature electronics is inexorably tied to the automotive sector. However, it is clear that the majority of automotive applications will be satisfied with incremental improvements to bulk silicon technology, with development dominated by cost considerations.

References

1. "World OEM Automotive Electronics to 2005," Freedonia Group, Market Study, June 2001, www.freedoniagroup.com.
2. Kobe, Gerry, "What's Driving the Growth," *Automotive Industries*, Vol. 180, No. 8, August 2000.
3. Hansen, Paul, "Current Trends: What are the roadblocks facing mass market, high-tech electronic content?," *Automotive Industries*, Vol. 181, No. 7, 26–29, July 2001.
4. Meyers, Bruce A., Jeff H. Burns, and Joseph M. Ratell, "Embedded Electronics in Electro-Mechanical Systems for Automotive Applications," *SAE International, Technical Paper Series*, #2001-01-0691, March 5–8, 2001.
5. "The Future of Automotive Electronics," *Automotive Engineering International*, October 2000.
6. Constapel, R., J. Freytag, P. Hille, V. Lauer, and W. Wondrak, "High Temperature Electronics for Automotive Applications," *Int. Conf. on Integrated Power Systems* (CIPS 2000), 20.-21.6.2000, Bremen, ETG Fachbericht 81, 46–52.
7. Wondrak, W., A. Boos, and R. Constapel, *Design for Reliability in Automotive Electronics, Part I: Semiconductor Devices*, Microtec, Hannover, 25–27, 299–302, 2000.
8. Lugert, G., E. Wirries, C. Beuther, *High Density-PCB Validation for Automotive Applications*, Electronic Forum, Waiblingen 2.-3.12., 1999 und Berlin: IEEE/ CPMT 28-29.6.1999, *Advances in PCB and Substrate Technologies*.
9. Lugert, G., T. Riepl, C. Beuther, E. Wirries, *Qualification of CSPs for Automotive Environment*, Flip Chip & Chip Scale Europe, Sindelfingen: Proceedings, March 14–15, 2000.
10. Riepl, T., G. Lugert, *Interconnection Materials for Flip Chip Technology in High Temperature Automotive Applications—A Comparative Study*, Micromat 2000, Berlin: Proceedings, 367–370, April 17–19, 2000.
11. Riepl, T., G. Lugert, R. Ingenbleek, W. Runge, L. Berchtold, *Integration of Micromechanic Sensors, Actuators and Miniaturized High Temperature Electronics in Advanced Transmission Systems*, Microtech 2000, Hannover: Proceedings pp. 599–604, September 25-27, 2000.
12. Lugert, G., S. Bolz, E. Wirries, T. Riepl, *Future Requirements and Technologies for Automotive Control Units*, in preparation.
13. Johnston, C., *Markets for High Temperature Electronics*, HITEN, 2000.

8

Smart Materials

Clifford M. Friend

CONENTS

Smart Context

Smart technologies, which encompass both smart materials and smart structures, are creating a sea change in engineering practice. Their fusion of conventional structural materials with aspects of information technology offers the prospect of engineering systems that can sense their local environment, interpret changes in this environment, and respond appropriately. This offers the possibility of engineering structures that can operate at the very limit of their performance envelopes and to their structural limits without fear of exceeding either; structures that can give maintenance engineers a full report on their performance history, as well as the location of defects, and structures that have the ability to counteract unwanted or potentially dangerous conditions, and even in the future, effect self-repair.

Smart technologies are currently under serious development in a range of sectors, including health and usage monitoring systems (HUMS) in aerospace and civil infrastructure, vibration control of buildings and pantographs on trains, and noise reduction systems in turboprop aircraft. They have even penetrated the sporting goods market, with skis incorporating adaptive vibration control, and the consumer product market with the launch of Russel Hobbs' Smart Thermocolor™ electric kettle.

Smart Technologies

The terms "smart structures" and "smart materials" are much used and abused. It is beyond the scope of this chapter to discuss the philosophical features of either term in any depth, however, it is important to take a clear view on the scope of technologies associated with this field. A useful starting point here is Culshaw's analysis, which concludes that single materials can never be construed as smart, since any single material can only "respond to external influences without any implicit or explicit information reduction potential. That is, the output from a single material can, at best, be a one-to-one function of an input stimulus, at worst it is a multi-valued function of the input stimuli. . . . Any claim for a material to be smart must then unavoidably necessitate the use of **hybrid materials** (this author's emphasis). . . . to provide the necessary adaptive functions." This is an important observation and implies that smartness can only be achieved by assembling material systems that create the required functionality. This instantly reveals a common abuse of the term smart material, which is often used to describe materials with unfamiliar transductions (for example, electric or magnetic field to strain, temperature to strain, etc.). Although clearly *not* smart by this definition, such materials have important functionality and can, and do, form important parts of smart material systems, as we shall see.

From Culshaw's viewpoint, the terms smart material systems and structures can be used interchangeably depending on the scale of material integration. Structures typically are used to describe systems close to conventional mechatronics (hybrid mechanical/electronic systems), and material systems are used when the scale of material integration is higher.

Smart systems can be created from a variety of technologies ranging from mechatronics using conventional approaches, such as accelerometers and servo-hydraulic actuators (Figure 8.1), to more integrated systems, which are currently the focus of much development work. Such highly integrated systems exploit a range of solid-state sensor and actuator technologies. These include sensual structures containing optical fiber (Figure 8.2) and piezo-ceramic/polymer strain sensors and adaptive systems containing novel piezo-ceramic, electro- and magnetostrictives, electro- and magnetorheological fluids (Figure 8.3), and shape–memory actuators (Figure 8.4). Magnetorheological fluids are discussed in detail in chapter 5.

Demonstrator projects exist world-wide, exploring the range of possible applications for technologies in sectors ranging from aerospace and civil

FIGURE 8.1
Hybrid mechatronic systems composed of accelerometers and actuators.

engineering to automobile and marine. A common feature of such programmes is their use of relatively sophisticated technologies; examples including the use of fiber optic techniques for sensing and actuation based on functional materials such as piezoceramics, electro- and magnetostrictives, and shape-memory alloys (SMAs). The use of such advanced technologies is symptomatic of the strong technology push that has dominated the development of this field.

FIGURE 8.2
Fiber-optic strain sensors based on interferometric or Bragg-grating technologies.

FIGURE 8.3
Magnetorheological fluids—a material with real-time adaptive rheology.

Technology Push or Application Pull?

There is no doubt that the development of smart technologies has been technology pushed. However, as development of these systems has increased, there is a greater need to match enabling technologies to the true requirements of a given application sector. This has not only focused on the appropriateness of certain advanced technologies for particular sectors—such as fiber optic sensing in the built environment, but also on the wider issue of whether smart technology is truly a collection of technologies, or instead a new paradigm for the design of products and systems. The latter is an

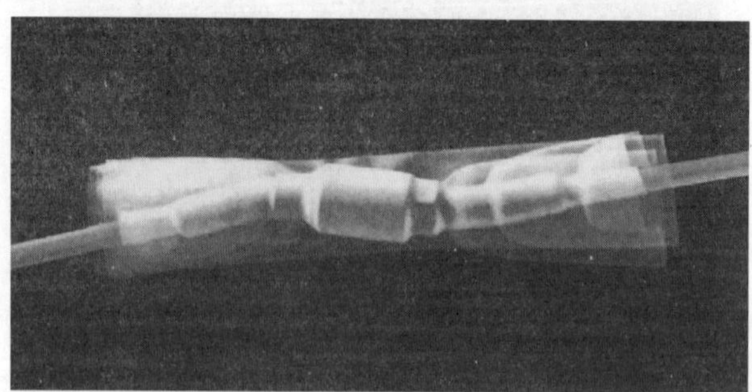

FIGURE 8.4
Shape-memory polymers—thermally driven solid-state actuators.

important issue since recognition that products can be adaptive offers a wide range of new and novel design options.

The issue of technology push has had a significant impact on the application of smart technologies within the automobile sector. Many smart applications have been identified, and even developed, in laboratory-scale demonstrations. These offer an interesting range of case studies on how appropriate technologies can be matched to the economics of this sector; where price of the technology, perhaps, dominates over the gain provided by the technology itself. Successful products will, therefore, be those with a strong market pull, where there is a good match between performance, technology, and price.

Smart Materials and the Automobile

Early Applications

In early smart automobile applications, researchers explored a range of products based on the mechatronic systems, which were at the time, increasingly dominating development programs. For example, exploiting fundamental work on the vibration control of large-scale structures, lateral thinking designers created applications, such as the wiperless windshield, where sensors detected rain, and piezoelectric actuators vibrated rain drops from the shield. Innovative, but often of limited utility, these systems, usually failed to transition to commercialization, although some elements of such systems (such as rain sensors) have eventually found application in automobiles.

At the same time, materials with unusual transductions, such as shape-memory alloys (SMAs), electro- and magnetorheological fluids, and photochromic glasses, became the focus of development programs for systems, such as active ride control, passive climate control through adaptive solar gain, and simple active control valves. From a technological viewpoint, these applications are interesting since they have proved technically feasible, and often a good match to applications. Indeed, a number of them, such as electrorheological and magnetorheological fluids and shape memory alloys have reached niche applications in, for example, vibration-controlled seats for off-road vehicles, and oil-flow valves in automatic gearboxes. However, these applications have been truly only niche. The use of shape memory alloy valves in Mercedes Benz's automatic gearboxes also illustrates another interesting feature of the application of such smart technologies. This is that the technology should not be seen in isolation, but instead as one solution to the specific requirements of the application. Here, a wide range of possible technological solutions—smart and non-smart—compete on price and utility. In the case of the gearbox application, the smart shape memory alloy component finally saw only limited use in one generation of product, due to a competing mechatronic solution that has continued in service. This tells one that smart technologies will break through only where there is an appropriate match between the application and technology, and where the smart solution has a unique selling proposition. A number of such applications are now emerging, and one can expect to see more as this new materials design approach develops further.

Well-Matched Applications

One well-matched automobile application is a return to vibration control, but no longer for unusual applications, such as windshields, but for active noise control. Here, the smart structural concept of sensors embedded within the structure of the automobile and active excitation of components, such as the roof panel, can be used to cancel noise within the passenger space. This is a mainstream smart-structural technology, transferred from developments in the turbo-prop aircraft sector, but appears to meet both the technological and commercial requirements of the automobile sector.

Perhaps more interesting has been the successful revisiting of technologies, such as shape memory alloys as actuators to replace electromechanical systems in components, such as climate control, and as an alternative to electrochromic rear-view mirrors. Both will shortly enter the market in Fiat cars where, through good matching between design and material selection, products such as shape memory alloy actuated electromechanical rear-view mirrors can be manufactured as a low-cost, but highly effective alternative to electrochromic technologies; doing to electrochromic materials what has often been done to shape memory alloys, i.e., producing a competing lower cost technological option for the same application.

Perhaps most interesting among these case studies is another Fiat concept that illustrates how the use of the smart adaptive design concept, selection of smart materials, and their integration into automobile structures offer the prospective of completely new automobile technologies of great significance. An example is the active bonnet/hood. Most automobiles will have great difficulties in meeting pedestrian safety legislation, which will come into force in Europe in the near future. The smart bonnet is based on other mainstream smart applications, but is one well-matched to the automobile sector, producing a system arising from a real and significant application pull. One solution to the challenge of optimizing an automobile for everyday use, including good aerodynamics, and that is safer to pedestrians during accidents, is to actively open the bonnet/hood during impact with a pedestrian. This alters the shape of the front of the automobile, minimizing injury to the pedestrian. The active control of such a smart bonnet, the sensing of impacts, and its functioning only during pedestrian impact—to minimize both false-positive and negative responses, and the selection of appropriate sensing technologies—such as piezoelectric polymers—is a major challenge that has only just begun, but illustrates how automobile performance, such as safety, could be improved in the future by the use of smart technologies.

Future

The automobile sector is only beginning to explore realistically the potential of smart technologies. In many cases, such applications are substitutional, replacing more complex subsystems with simpler smart solutions. However, it is clear from concepts, such as the smart bonnet/hood, that designs are already looking beyond substitutional applications to novel systems, where the required functionality is *only* deliverable through the smart route. The impact of such step-change designs are often seen in purely engineering terms, however, the use of smart design is also likely to extend beyond the conventional engineering domain. For example, the penetration of smart materials systems into the consumer goods market has already opened new applications at the interface with industrial design. This area is of direct relevance to cockpit design within automobiles from both the ergonomic and industrial design/styling viewpoints. Recent work on smart seats for aircraft also points to other interesting user-focused applications. The development of new smart automobile applications, therefore, appears to be constrained only by the mind-set of engineering and industrial designers.

Many of the materials used to construct smart systems are already available off-the-shelf but few of these functional materials have ever been developed for smart engineering applications—many falling into the familiar category of "interesting materials looking for an application." A future driver in this field will, therefore, be the optimization of such materials for integration into smart products and, where necessary, their further development to meet the particular performance requirements of different families of applications. For example, in automobile applications, under-hood subsystems could be considerably simplified by the use of solid-state actuators, such as shape memory alloys, if the operating temperatures of these alloys could be raised. So, not only is there a future challenge to engineering and industrial designers, but also to materials engineers and scientists, to ensure that the enabling materials technologies themselves are optimized to meet the strong market pull for smart automobile solutions.

Summary

Smart technologies are creating a sea change in engineering practice that will not leave the automobile sector unaffected. The fusion of conventional structural materials with aspects of control offers the prospect of a new design paradigm for automobile materials and structures. To exploit such opportunities, we must avoid the attractions of pushing technologies, but instead take a holistic design approach, responding to market pull by developing smart technological solutions that best match the needs of the application on the basis of technology, cost, and the unique solutions offered by smart technologies. Such effective matching of smart technologies to automobile requirements will then offer new solutions in applications ranging from under-hood and passenger comfort subsystems, to smart noise and active structural control.

Section 3

Light Metals

Automotive engineering technology roadmaps from around the globe are unanimous in their call for new materials to allow for the reduction in automotive weight. Reduced weight improves fuel efficiency and so cuts emissions. The challenge to ferrous materials by aluminium alloys in structural applications and body panels has spurred the development of a new generation of commodity high-strength low alloy and other steels. Nonetheless, the fraction of aluminium alloys in automobiles has continued to rise steadily, along with the increasing penetration of magnesium alloys in niche applications. This section considers the remaining challenges that must be met if light alloys are to continue to find increasing use in automotive structures, displacing ferrous-based materials, as well as the increasing competition from lightweight polymeric-based composites. Key areas for future technological focus are identified as:

- Formability of aluminium and magnesium based alloys
- Crashworthiness
- Lightweight advanced materials (foams and cored materials)

9

Formability of Aluminum Alloys

Hirofumi Inoue

CONTENTS

Introduction

In recent years, lightening the weight of automobiles is being carried out to reduce fuel consumption as an answer to the earth-environmental problem. Above all, the adoption of aluminum alloy sheets to automotive body panels has been attempted actively in view of recycling of the materials.[1] When aluminum alloys are actually used for auto body panels, a measure of evaluating press formability is necessary, because it is an important property together with strength and ductility. Deep drawability in low carbon steel is greatly influenced by the Lankford value,[2] i.e., the r-value. Aluminum alloy sheets are also considered to show a positive correlation between the limiting drawing ratio (LDR) and the r-value under certain conditions[3] but the r-value is not necessarily used as an effective measure of evaluating deep draw-ability.[4] In the case of aluminum alloys, it is possible to express deep drawability by using the average and minimum of the r-values in the sheet, $\bar{r}$ ($\bar{r} = (r_0 + 2r_{45} + r_{90})/4$) and r_{min}, and the average of the n-value (work-hardening exponent), $\bar{n}$[3]. Accordingly, planar anisotropy of the r-value in sheet materials must be investigated in detail to develop aluminum alloy sheets with excellent deep drawability.

TABLE 9.1

Tensile Properties and Formability of Aluminium Alloys for Autobody Panels with Sheet Gauge of 1 mm[5]

Alloy and Temper	T. S. (MPa)	Y. S. (MPa)	Elongation (%)	n-Value	r-Value	Olsen Cup Height (mm)
2002-T4	330	180	26	0.25	0.63	9.6
2036-T4	340	195	24	0.23	0.75	9.1
2037-T4	310	170	25	0.24	0.7	9.4
2038-T4	325	170	25	0.26	0.75	—
5182-O	275	130	26	0.33	0.8	9.9
5182-SSF	270	125	24	0.31	0.67	9.7
6009-T4	230	125	25	0.23	0.7	9.7
6010-T4	290	120	24	0.22	0.7	9.1
6111-T4	290	160	27	—	—	8.4
6016-T4	235	125	28	0.26	0.7	—
6022-T4	255	152	26	0.25	0.67	—
SPCC (steel)	315	175	42	0.23	1.39	11.9

Since the r-value is closely related to crystallographic orientation in polycrystalline materials, it can be improved through texture control. However, in annealed aluminum alloy sheets produced by conventional processes, the average r-value generally exhibits low values (less than 1.0), and it is in the range of 0.6 to 0.8 for 2xxx, 5xxx, and 6xxx series aluminum alloys for auto body panels, as shown in Table 9.1.[5] This is about half of the average r-value of low carbon steel. In order to find a possibility of improvement of the r-value, the relation between texture and the r-value must be exactly grasped by quantitative texture analysis using crystallite orientation distribution function (ODF).[6]

This chapter describes texture in aluminum alloy sheets, relations between texture and r-value, and the correlation between the r-value and the limiting drawing ratio. In addition, methods for improving formability of aluminum alloy sheets are discussed in terms of texture control.

Texture of Aluminum Alloy Sheets

Aluminum alloys used as auto body panels are mainly 5xxx and 6xxx series, which have relatively excellent formability and corrosion resistance.[5] These alloy sheets are usually employed in an annealed (O) or solution heat-treated (T4) state after cold rolling. Since recrystallization occurs during the heat treatment, recrystallization texture is related to the r-value of the sheets. However, rolling texture that strongly affects the formation of recrystallization texture is also important in discussing optimum processing for texture control. An example of recrystallization texture for 5052 and 6061 alloys[7] is shown in Figure 9.1. Both alloys before annealing had similar rolling textures consisting mainly of the β-fiber of {112} <111> (Cu-orientation) – {123} <634>

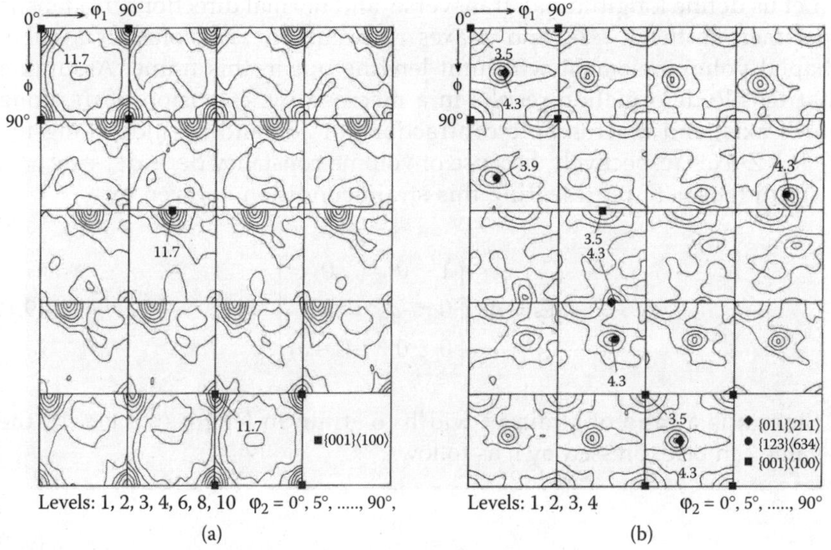

Levels: 1, 2, 3, 4, 6, 8, 10 $\varphi_2 = 0°, 5°, ..., 90°,$

(a)

Levels: 1, 2, 3, 4 $\varphi_2 = 0°, 5°, ..., 90°$

(b)

FIGURE 9.1

Orientation distribution functions of (a) 5052 and (b) 6061 aluminum alloy sheets cold rolled to 90% reduction and subsequently annealed for 3 ks at 573 and 623 K, respectively.

(S-orientation) – {011} <211> (Bs-orientation). Nevertheless, the recrystallization textures are significantly different in preferred orientations. A 5052 alloy shows cube texture of {001} <100>, while a 6061 alloy shows retained β-fiber and cube components. The retained β-fiber is composed of {123} <634> and {011} <211> orientations contained in rolling texture, and indicates higher orientation density than cube orientation. From the facts that softening by recrystallization occurs more slowly in a 6061 alloy than in a 5052 alloy and a number of pancake-like grains exist in the recrystallized microstructure of a 6061 alloy, it is speculated that fine particles present before annealing, and particles precipitated during annealing, inhibit the migration of subgrain boundaries and recrystallized grain fronts, and consequently, continuous recrystallization occurs in part. Fine precipitates thus play an important role in the evolution of recrystallization texture.

Relation between Texture and *r*-Value

Prediction of the *r*-value from experimentally measured texture should be performed to clarify the relation between texture and the *r*-value. Bunge[6,8] proposed a method for predicting the *r*-value from an orientation distribution function in polycrystalline materials based on the Taylor theory[9] and calculated planar anisotropy of the *r*-value for a low carbon steel sheet.[10] This method is applied to aluminum alloy sheets.

Let us define longitudinal, transverse, and normal directions in a tensile specimen as the X-, Y-, and Z-axes respectively, and consider a cube-shaped volume element with unit lengths put in this frame. Assuming that tensile deformation results in a macroscopic extension of $d\varepsilon$ along the X-axis, and macroscopic contractions of $qd\varepsilon$ and $(1-q)d\varepsilon$ along the Y- and Z-axes respectively, because of volume constancy $d\varepsilon_x + d\varepsilon_y + d\varepsilon_z = 0$, a strain tensor E_s representing this strain condition is given by

$$E_s = d\varepsilon \cdot \begin{bmatrix} 1 & 0 & 0 \\ 0 & -q & 0 \\ 0 & 0 & -(1-q) \end{bmatrix} \tag{9.1}$$

Since q is a ratio of strain in width to strain in length ($0 \le q \le 1$), the r-value can be expressed by q as follows:

$$r = \frac{q}{1-q} \tag{9.2}$$

Here, we adopt the Taylor model, i.e. an assumption that individual grains in a polycrystal undergo the same homogeneous strain as macroscopic strain shown in Equation 9.1 to satisfy the continuity of strain at grain boundaries. With respect to a certain value of q, we calculate the Taylor factor M for orientations g present in the Euler orientation space (φ_1, Φ, φ_2) at regular angular intervals. To determine the Taylor factor of a textured polycrystalline material, an average of M over the whole orientation space weighted by an orientation distribution function $f(g)$ is taken by the following equation.

$$\bar{M}(q) = \oint M(q,g)f(g)\,dg \tag{9.3}$$

A plot of $\bar{M}(q)$ against q almost always gives a concave curve with the minimum value in the range of $q = 0$ to 1. Thus, the value of q at a minimum $\bar{M}(q)$, namely q_{min} is determined in the same way as a method by Hosford and Backofen,[11] and then the r-value is obtained from q_{min} by Equation 9.2.

Figure 9.2 shows planar anisotropy of the r-value measured experimentally and calculated from orientation distribution function shown in Figure 9.1.[7] Curves of measured and calculated values are qualitatively similar to each other for both alloys. Particularly, these values are numerically consistent for a 6061 alloy. For a 5052 alloy consisting mainly of cube texture, the calculated r-value, however, indicates a rather lower value at a direction of 45° to the rolling direction than the measured one. This is not essentially due to the inhomogeneity in texture through thickness and the textural change

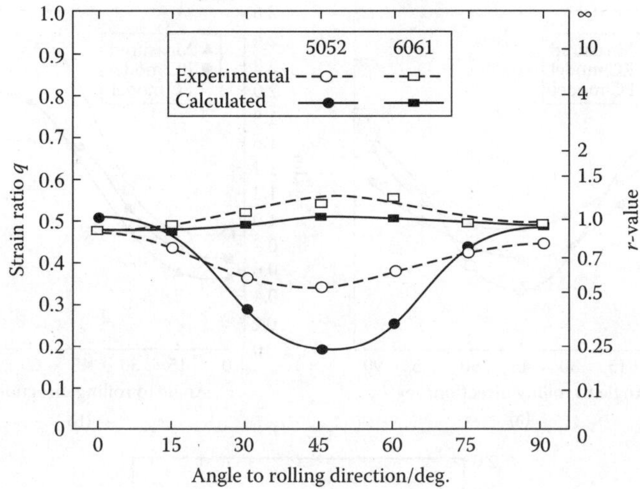

FIGURE 9.2

Measured and calculated r-values for annealed 5052 and 6061 aluminum alloy sheets as a function of angle to rolling direction.

during tensile deformation, but is probably due to the inadequacy of a model used for calculation. Since locally inhomogeneous deformation tends to take place actually in Al-Mg alloys, such as a 5052 alloy, the conventional Taylor model supposing homogeneous strain, namely the full constraints model (FC model)[9] seems to be rather unsuitable for prediction of the r-value in a 5052 alloy.

Therefore, it is necessary to calculate the r-value by the relaxed constraints model (RC model)[12] that relaxes some shear strains from a constrained state with shear strain components of zero in Equation 9.1. The authors predicted the r-values in ultra-low carbon steel sheets using the relaxed constraints models with various relaxed shear components, and reported that the relaxation of constraint in a shear strain component γ_{23} perpendicular to the tensile direction was necessary for exact estimation of the r-values.[13] Here, the subscripts 2 and 3 mean transverse and normal directions in a tensile specimen, respectively. In aluminum alloys as well as steel, the relaxed constraints model on the γ_{23} component has been employed to obtain better agreement between measured and calculated values, provided that γ_{23} is not set quite free, but $|\gamma_{23}|$ is set between 0 and 0.5. Some results in annealed 5052 and 6061 alloys are shown in Figure 9.3. For a 5052 alloy, the r-values calculated by the relaxed constraints model are, on the whole, closer to the measured r-values than those by the full constraints model, whereas for a 6061 alloy the r-values calculated by the full constraints model are closer to the measured r-values. Thus, the r-values of 5xxx series Al-Mg alloys, which are of the solid solution strengthening type, may be successfully estimated by the

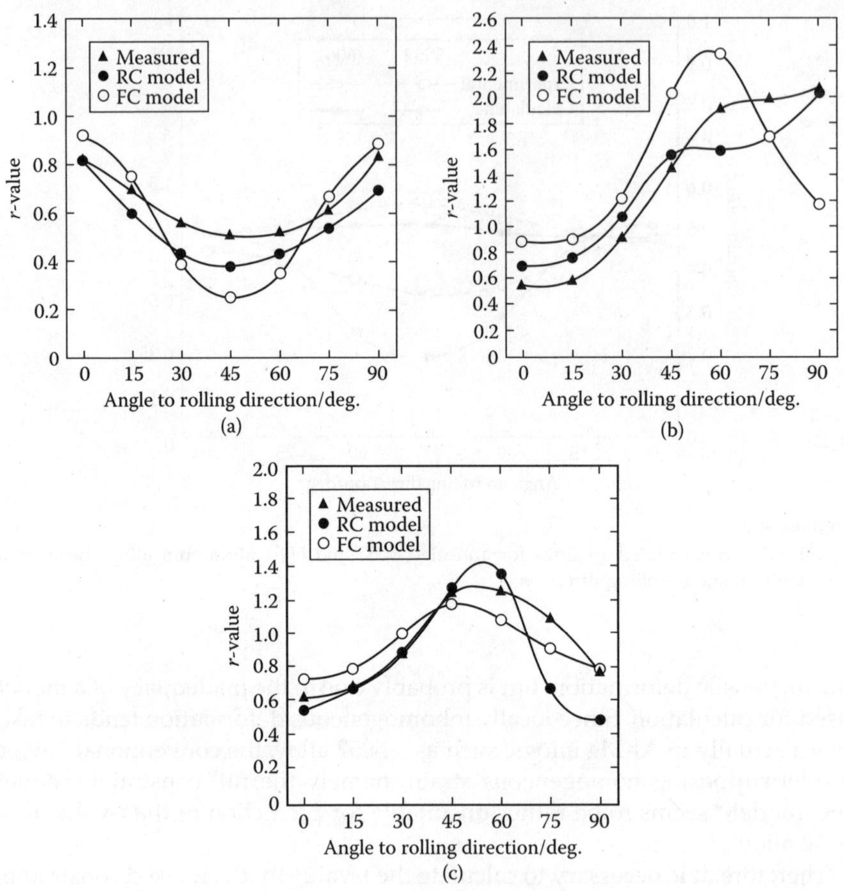

FIGURE 9.3

Comparisons between *r*-values calculated by the relaxed constraints (RC) and full constraints (FC) models for aluminum alloy sheets cold rolled to 90% reduction and annealed under different conditions. (a) 5052 alloy annealed at 573 K for 500 s, (b) 5052 alloy annealed at 543 K for 1 ks, and (c) 6061 alloy annealed at 623 K for 500 s.

relaxed constraints model on shear strain γ_{23} rather than the full constraints model, because they tend to be deformed inhomogeneously.

In order to reveal what orientations are effective in improving the *r*-value, planar anisotropy of the *r*-value for various ideal orientations with Gaussian distribution[6] has been estimated using the full constraints model.[7] Figure 9.4 shows the calculated results for representative orientations observed in rolled and recrystallized aluminum alloy sheets. A {111} <110> orientation observed occasionally in the surface region of a sheet possesses excellent *r*-values at any direction. It is, however, difficult to develop this orientation through sheet thickness. For cube orientation {001} <100>, the *r*-value is

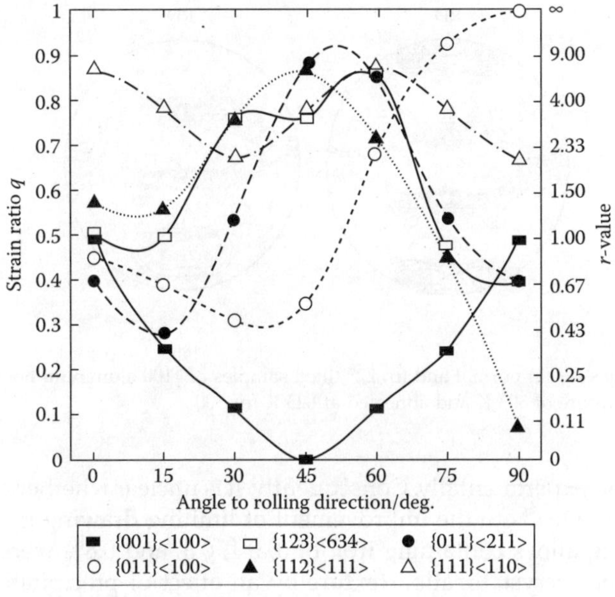

FIGURE 9.4

Calculated *r*-values for typical ideal orientations with Gaussian distribution of $\phi_0 = 10°$ as a function of angle to rolling direction.

approximately 1.0 at directions of 0° and 90°, but indicates fairly low values at other angles, especially zero at a direction of 45°. On the contrary, the *r*-value of R-orientation {123} <634> is about 4 in the range of 30° to 60°, and is not so low even at 0° and 90°. Bs- and Cu-orientations as well as R-orientation indicate fairly high *r*-values in the vicinity of 45°. From this figure, it is obvious that retained rolling texture components such as {123} <634> are effective for obtaining high *r*-values by realizable orientations.

Correlation between *r*-Value and Limiting Drawing Ration

The influence of the *r*-value and *n*-value on limiting drawing ratio (LDR) has been examined experimentally and theoretically[3,14] limiting drawing ration somewhat increases with increasing *n*-value when the $\bar{r}$-value is constant. Since the *n*-value does not change so much for annealed aluminum alloys, there is slight influence of the *n*-value on limiting drawing ratio. On the other hand, limiting drawing ratio seems to increase with increasing $\bar{r}$-value for aluminum alloys. As described at the beginning, the $\bar{r}$-value of usual aluminum alloys is in a narrow range from 0.5 to 1.0, so that it is difficult to find a positive correlation between the $\bar{r}$-value and limiting

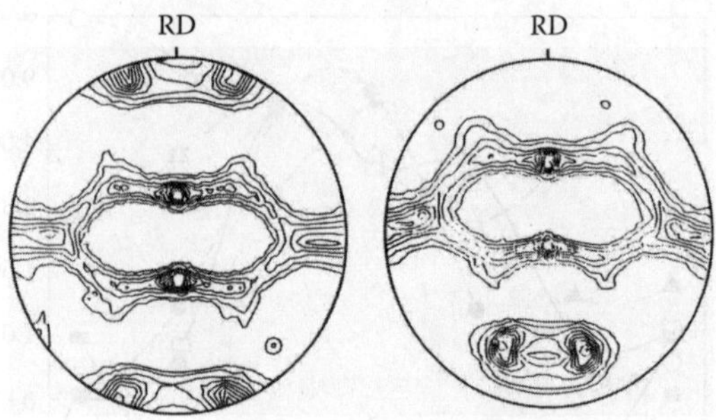

FIGURE 9.5
{111} pole figures for (a) normal and (b) 22° tilted samples of 1100 aluminum hot rolled at the starting temperature of 803 K and annealed at 623 K for 300 s.

drawing ratio experimentally. Consequently, it is unclear whether an increase of the $\bar{r}$-value leads to the improvement of limiting drawing ratio.

Al-4.3% Mg alloys containing iron of 0, 0.1, 0.3, and 0.8% were prepared to change the recrystallization texture by an effect of precipitation during annealing. A hot rolled 1100 aluminum plate of 11 mm in thickness was also prepared in order to obtain a high $\bar{r}$-value by cutting out sheet samples at an oblique plane inclined by 22° from the rolling plane toward the rolling direction. Recrystallization texture of the sheet samples contains a near {111} orientation as a main component, because the hot rolled plate before cutting had a strong β-fiber texture that shows strong {111} pole density at the position inclined by 22° from the normal direction toward the rolling direction (Figure 9.5). As a result, they showed a considerably high $\bar{r}$-value of 1.61. Limiting drawing ratio of such materials was measured using flat-headed punch. A correlation between the limiting drawing ratio and the $\bar{r}$-value is shown in Figure 9.6. It is obvious from this figure that the limiting drawing ratio increases linearly with increasing $\bar{r}$-value. This means that

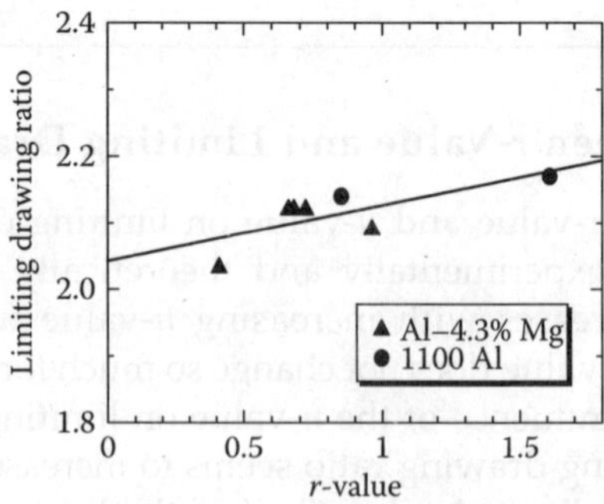

FIGURE 9.6
Correlation between limiting drawing ratio and average r-value.

the improvement of deep drawability is possible in aluminum alloys if the texture is adequately controlled.

Methods of Improving Formability by Texture Control

From the above section, it is found that increasing the $\bar{r}$-value by texture control is effective in the improvement of formability. Here, guidelines of texture control are discussed on the basis of relations between preferred orientation and the r-value shown in Figure 9.4. An important point of texture control is what orientations should be developed preferentially from various orientations reported actually in aluminum alloys.

There are two different guidelines to obtain a high $\bar{r}$-value. In analogy with steel, {111} <*uvw*> orientations such as {111} <110> and {111} <112> have high $\bar{r}$-values and small planar anisotropy, but it is difficult to develop these orientations throughout thickness by conventional rolling and annealing processes. Recently, there are many studies on the formation of shear texture by warm rolling and differential speed rolling (asymmetric rolling). Shear texture is attributable to additional shear deformation and consists mainly of a <110>//RD fiber of {001} <110> – {112} <110> – {111} <110>. If {111} <110> can be developed after annealing, a high $\bar{r}$-value will be obtained. However, since the shear texture is weakened by annealing and the <110>//RD orientations other than {111} <110> also remain after annealing, the $\bar{r}$-value is not so high as expected ($\bar{r} = \sim1.0$).[15] It is necessary to retain a strong shear texture including {111} <110> even after recrystallization. Another way to obtain a high $\bar{r}$-value is as follows. For raising the r-value at nearly all directions in a sheet plane, R-orientation {123} <634> is considered to be most effective from Figure 9.4, but this orientation shows a relatively low r-value of about 0.7 at the 90° direction. Therefore, it is required that Goss orientation {011} <100>, which raises the r-value at the 90° direction, is developed to some extent. Actually, Goss orientation has been observed in some examples.[16] The selection of alloy compositions, annealing conditions, and so on causing moderate competition between precipitation and recrystallization, will be necessary to develop R-orientation as a main component and Goss orientation as a secondary component. At least {001} <*uv*0> textures, such as cube orientation, are unfavorable for improving the r-value.

Warm rolling was actually carried out for the above Al–4.3% Mg alloys to obtain a high $\bar{r}$-value by the formation of shear texture including {111} <110>. Warm rolled sheets exhibited a sharp shear texture consisting of <110>//RD fiber components such as {001} <110>, {112} <110>, and {111} <110> in a range from the surface to the quarter thickness, and a sharp β-fiber texture in a range from the quarter thickness to the central layer.[17] The shear texture with a main component of {111} <110> remains after recrystallization annealing, as

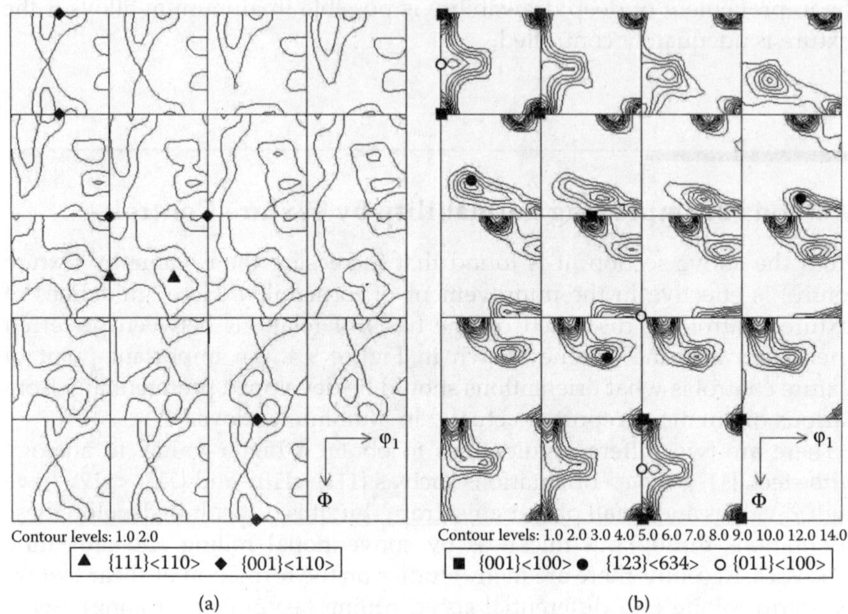

FIGURE 9.7
Orientation drawing functions in (a) surface and (b) central regions of Al-4.3% Mg-0.3% Fe alloy sheet warm rolled at 623 K to 87% reduction and subsequently annealed at 573 K for 10 ks.

shown in Figure 9.7, although the orientation density is fairly low compared with the as-rolled state. In the central region, R-orientation {123} <634> and Goss orientation {011} <100> coexist with a sharp cube orientation {001} <100>. Texture components other than {001} <110> and cube orientations in the surface and central layers seem to be suitable for improving the r-value, as explained above. The r-value of a sample with such texture was measured together with those of a sample with cube and R-orientations obtained by conventional cold rolling and annealing. Figure 9.8 shows a comparison between the r-values of warm- and cold-rolled sheets annealed under the identical condition. Obviously, the warm-rolled sheet exhibits higher r-values than the cold-rolled sheet in every direction. Texture control through warm rolling is thus effective in the improvement of the $\bar{r}$-value closely related to the LDR, that is, deep drawability.

Summary

The effect of crystallographic texture on the formability of aluminum alloy sheets, i.e., the relations between texture and r-value, and the correlation between the r-value and limiting drawing ratio has been described. Planar

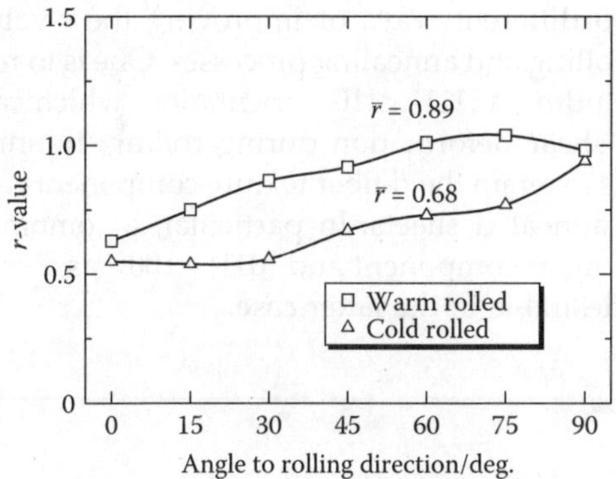

FIGURE 9.8
Measured r-values for Al-4.3% Mg-0.3% Fe alloy sheets annealed at 573 K for 10 ks after cold and warm rolling.

anisotropy of r-value in a sheet is quantitatively predicted from texture on the basis of the Taylor model. The full constraints model is shown to be suitable for calculating the r-value of 6xxx series aluminum alloys, whereas the relaxed constraints model is suitable for calculating that of 5xxx series alloys. The measured results of limiting drawing ratio and average r-value for materials with different textures showed a positive correlation between them. This fact means that the improvement of formability by texture control is possible in aluminum alloy sheets. Methods of improving formability are proposed based on the predicted results of the r-value for various ideal orientations. Texture control through warm rolling actually leads to the improved average r-value due to the formation of shear texture beneath the surface of a sheet.

Some important conclusions in this chapters are as follows:

Recrystallization textures of 5052 and 6061 aluminum alloy sheets are significantly different in spite of the evolution of similar rolling textures. This is due to the effect of precipitation during annealing in a 6061 alloy.

A suitable model for predicting the r-value from texture depends on alloy composition. For a 6061 alloy, the full constraints model leads to good agreement between calculated and measured r-values, while for a 5052 alloy, the relaxed constraints model gives better agreement between them than the full constraints model. This is related to the fact that inhomogeneous deformation tends to take place in 5xxx series alloys.

There is a positive correlation between limiting drawing ratio and average r-value for aluminum and its alloys, as well as for steel. Consequently, the improvement of deep drawability is possible in aluminum alloys by appropriate texture control.

There are two different ways of improving the r-value by texture control in rolling and annealing processes. One is to retain the shear texture including a {111} <110> orientation, which can develop by additional shear deformation during rolling, in annealed sheets. The other is to retain the β-fiber texture components formed during rolling, in annealed sheets. In particular, a combination of {123} <634> as a main component and {011} <100> as a secondary component is desirable in the latter case.

References

1. Abe, Y., M. Yoshida, O. Noguchi, M. Matsuo, and T. Komatsubara, *J. Japan Soc. Tech. Plasticity*, Vol. 33, 365, 1992.
2. Lankford, W. T., S. C. Snyder, and J. A. Bausher, *Trans. ASM*, Vol. 42, 1197, 1950.
3. *Formability of Aluminum Alloy Sheets*, Edited by Metal Forming Section, Japan Institute of Light Metals, Tokyo, 30, 1985.
4. Tozawa, Y., *J. Japan Soc. Tech. Plasticity*, Vol. 33, 782, 1992.
5. Uno, T., *Sumitomo Light Metal Technical Reports*, Vol. 42, 100, 2001.
6. Bunge, H. J., *Texture Analysis in Materials Science*, Butterworths, London, 1982.
7. Inoue, H., and N. Inakazu, *J. Japan Inst. Light Met.*, Vol. 44, 97, 1994.
8. Bunge, H. J., *Kristall und Technik*, Vol. 5, 145, 1970.
9. Taylor, G. I., *J. Inst. Met.*, Vol. 62, 307, 1938.
10. Bunge, H. J., and W. T. Roberts, *J. Appl. Cryst.*, Vol. 2, 116, 1969.
11. Hosford, W. F., and W. A. Backofen, *Fundamentals of Deformation Processing*, Syracuse University Press, 259, 1964.
12. Honneff, H., and H. Mecking, *Proc. 5th Int. Conf. on Textures of Materials*, Edited by G. Gottstein and K. Lücke, Springer-Verlag, Berlin, Vol. I, 265, 1978.
13. Inoue, H., and T. Hasegawa, *Proc. 1998 Japanese Spring Conf. for Technology of Plasticity*, Japan Soc. Tech. Plasticity, Tokyo, 223, 1998.
14. Logan, R. W., D. J. Meuleman, and W. F. Hosford, *Formability and Metallurgical Structure*, Edited by A. K. Sachdev and J. D. Embury, The Metallurgical Society, 159, 1987.
15. Sakai, T., H. Inagaki, and Y. Saito, *Proc. 12th Int. Conf. on Textures of Materials*, Edited by J. A. Szpunar, NRC Research Press, Ottawa, 1142, 1999.
16. Ito, K., *J. Japan Inst. Light Met.*, Vol. 43, 285, 1993.
17. Inoue, H., and T. Takasugi, *Z. Metallkd.*, Vol. 92, 82, 2001.

10

Ductile Magnesium

Toshiji Mukai and Kenji Higashi

CONTENTS

Introduction

Recently, a number of magnesium alloys have been used as structural components, such as cases for portable electronic equipment due to their high specific strength and good thermal conductivity. To reduce the weight of automobiles to reduce energy consumption, magnesium alloys are also expected to be used in structural components. In order for magnesium to be used in structural components, it is necessary that it exhibits sufficient strength and ductility under dynamic loading as well as static loading at ambient temperature. Elongation-to-failure at a dynamic strain rate compared to a static strain rate for some structural metals[1-3] are shown in Figure 10.1. Elongation-to-failure of magnesium at a dynamic strain rate is lower than that at a static strain rate[1] as well as other hexagonal close parted metals, such as zinc.[2] Therefore, the microstructure must be improved before magnesium can be used as a structural material. Wilson[4] demonstrated that the ductility of pure magnesium could be enhanced by refining its grain size at a static strain rate. Mohri et al.[5] has also reported the ductility enhancement of an Mg-Y-RE (rare earth) alloy by hot extrusion. Mukai et al.[6] investigated the ductility enhancement for the same alloy under dynamic loading, stating that the enhancement of ductility was due to refinement of the

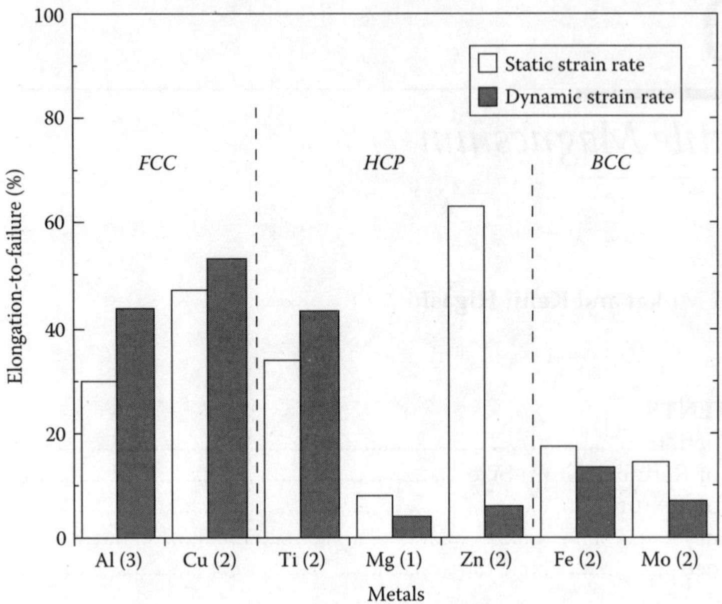

FIGURE 10.1
Elongation-to-failure at a dynamic strain rate comparing with that at a static strain rate for some structural metals.

microstructure. Thus, control of the grain structure raises the possibility for the development of structural magnesium alloys with high ductility at dynamic strain rates. In this study, effects of grain size refining, and controlling texture on the strength and ductility of magnesium alloys are investigated at dynamic strain rates.

Effect of Refining Grain Size

The alloy used in the present study is a commercial magnesium alloy of ZK60 (Mg-6Zn-0.5Zr, by wt. %). The material was received as an extruded bar. In this study, ZK60 was annealed at 773 K for 15 minutes (designated as ZK60-FG) and for 8 hours (ZK60-CG). The average grain size of ZK60-FG and ZK60-CG were measured to be 4 µm and 120 µm, respectively. Tensile specimens, machined directly from the extruded bars, had their tensile axis parallel to the extruded direction. Tensile tests at a dynamic strain rate ($\sim 10^3 s^{-1}$)

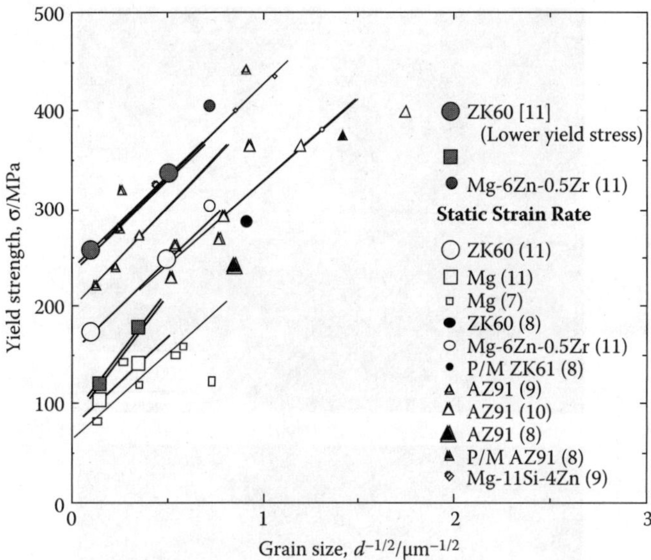

FIGURE 10.2

Hall–Petch relations of ZK60 magnesium alloy at a dynamic and static strain rate. Included are data of some magnesium alloys at a static strain rate.[7–11]

were performed using a modified Hopkinson pressure bar. All tests were conducted at room temperature.

The effect of grain refinement on the yield strength, i.e., the Hall–Petch (H–P) relations in the present materials are shown in Figure 10.2 for static and dynamic strain rates. Included are data for magnesium[7] and some magnesium alloys.[8–11] For the result of the present ZK60 alloy obtained at a dynamic strain rate, the stress was taken as the lower yield stress. The Hall-Petch slope is similar to those for other magnesium and magnesium alloys. All materials in Figure 10.2 were fabricated by extrusion and exhibit similar slope. Therefore, it is suggested that the developed texture may be similar. The yield strength of ZK60 is noted to be increased with refining grain size at the dynamic strain rate. The data at the dynamic strain rate are also represented by a single line. It is well known that the slope of the line, k-value, depends on the slip system.[12] Since the line at the dynamic strain rate is almost parallel to that at the quasi-static strain rate, it is noted that the effect of refining grain structure on the flow stress is essentially similar for both strain rates.

The elongation-to-failure of ZK60-FG is measured to be 25% and is obviously larger than that of ZK60-CG (16%) under the dynamic tensile loading. Inspection of fractured specimen revealed that the fracture of the coarse-grained ZK60 was caused by a coalescence of macroscopic cracks along the twin boundary and grain boundary (Figure 10.3a), while the macroscopic cracks were not observed in the fine-grained alloy due to the absence of macroscopic twin boundaries (Figure 10.3b).[11]

FIGURE 10.3
Deformed microstructure of ZK60 magnesium alloy with different grain sizes of (a) 120 μm and (b) 4 μm. White arrows indicate the evidence of macro-cracking.

Effect of Texture

It has been reported that an extruded AZ80 magnesium alloy exhibited a strong texture: the majority of basal-plane arranged parallel to the extruded direction.[13] The deformation perpendicular to the extrusion direction is limited and the alloy is fractured without macroscopic necking. Thus, one of the possible procedures for the ductility enhancement can be modification of structure with the rearrangement of the distribution of basal plane in a wrought magnesium alloy. In the present study, equal channel angular extrusion (ECAE) is selected for the experimental wrought process. It has been proposed that equal channel angular extrusion was a unique technique to apply heavy shear strain to the material, and developed the ultra-fine-grained

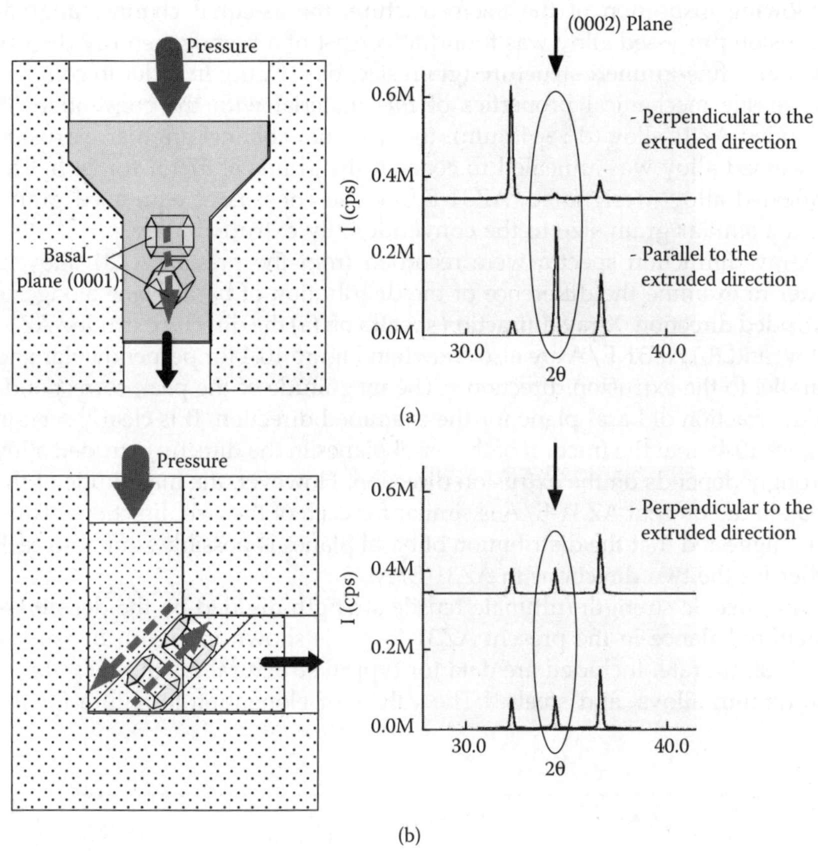

FIGURE 10.4
Schematic illustration and X-ray diffraction spectra of (a) the directly extruded AZ31 alloy and (b) AZ31-E/A.

structure for several kinds of metallic materials.[14] Very limited data are currently available for equal channel angular extrusion processed magnesium.[15] Figure 10.4 illustrates schematically the processes of (a) the conventional direct extrusion, and (b) equal channel angular extrusion.

In the magnesium processed by direct extrusion, it has been reported that the basal plane in the majority of grains is arranged along the extruded direction, as shown in Figure 10.4a. On the other hand, equal channel angular extrusion processes can apply a simple shear to the material at the 90 degree angled channel portion, as shown in Figure 10.4b. The basal planes in the majority of grains rearrange in the shearing direction. In this study, equal channel angular extrusion was conducted on AZ31(Mg-3Al-1Zn-0.2 Mn, by wt. %) by a die with a 90 degree angled channel and at a temperature of 473 K. Equal channel angular extrusion was conducted by 8 passages through the die with rotation of the extruded bar +90° for the repetitive extrusion.

Following inspection of the microstructure, the as-equal channel angular extrusion processed alloy was found to consist of a homogeneously distributed and fine-grained structure (grain size, d: ~ 1 μm). In order to compare the tensile mechanical properties of this material with the conventionally extruded AZ31 alloy (d : ~ 15 μm), the as-equal channel angular extrusion processed alloy was annealed to coarsen the grains at 573 K for 24 h. The annealed alloy (designated, AZ31-E/A) also consists of equi-axed grains with a similar grain size to the conventionally extruded alloy.

X-ray diffraction spectra were recorded from the present AZ31 alloy in order to examine the difference of the distribution of basal plane along the extruded direction. X-ray diffraction spectra of: (a) the directly extruded AZ31 alloy; and (b) AZ31-E/A are also shown in Figure 10.4 for perpendicular and parallel to the extrusion direction.[16] The magnitude of the peak corresponds to the fraction of basal plane for the examined direction. It is clearly seen in Figure 10.4a that the fraction of the basal planes in the directly extruded alloy strongly depends on the extrusion direction. However, the magnitude of the peak in the present AZ31-E/A is similar for each of the two directions. Thus, it is suggested that the distribution of basal planes is possibly similar to each other for the two directions in AZ31-E/A.

The specific strength (ultimate tensile strength divided by the density)—ductility balance in the present AZ31-E/A—is shown in Figure 10.5 at the static strain rate. Included are data for typical structural magnesium alloys, aluminium alloys, and steels.[17] The values of elongation-to-failure for the

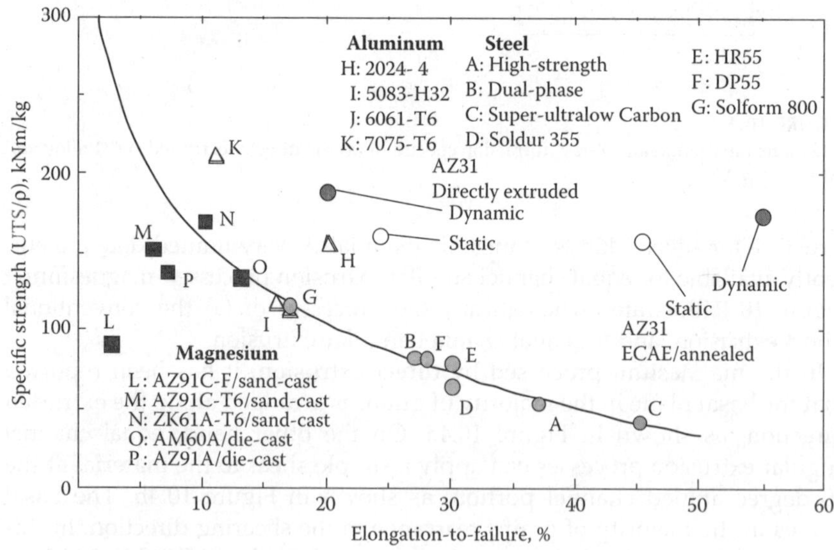

FIGURE 10.5
Specific strength–ductility balance of a wrought AZ31 magnesium alloy. Included are data of selected alloys of cast-magnesium, aluminum, and steels.[17]

magnesium alloys are markedly lower than those of structural steels. However, AZ31-E/A exhibits a similar value of specific strength compared with the cast magnesium alloys, and a larger value of elongation-to-failure than steels. Tensile tests at a dynamic strain rate revealed that the texture controlled alloy exhibited a similar ductility under dynamic loading. Thus, it is noted that the present alloy has a high possibility for structural use under dynamic loading.

Summary

Improvement of tensile mechanical properties under dynamic loading was demonstrated for commercial magnesium alloys. The yield stress of the ZK60 alloy increased with refining grain-size at the dynamic strain rate with a similar Hall–Petch slope to that at a quasi-static strain rate. Enhancement of ductility can be also achieved by refining the grain structure for ZK60 alloy. Simple shear was applied to a commercial AZ31 magnesium alloy by equal-channel-angular-extrusion for the development of a different texture compared with the conventionally extruded alloy. X-ray diffraction spectra examined for the parallel and perpendicular section to the extrusion direction suggested that the fraction of basal plane was similar to each other for equal channel angular extrusion/annealed alloy, while it was obviously different for the conventionally extruded alloy. It was found from the tensile test that the elongation-to-failure of the equal channel angular extrusion/ annealed alloy exhibited ~ 50%, which was twice larger than that of the conventionally extruded alloy.

References

1. Mukai, T., et al., *Light Weight Alloys for Aerospace Application III*, (TMS, Warrendale, 483–488, 1995.
2. Kawata, K., et al., *Proc. Symp.* HDP IUTAM, 313–317, 1968.
3. Mukai, T., K. Higashi, and S. Tanimura, *Mater. Sci. Eng.*, A176, 181–186, 1994.
4. Wilson, D .V., *J. Inst. Metals*, Vol. 98, 133–143, 1970.
5. Mohri, T., et al., *Mater. Sci. Eng.*, A257, 287–294, 1998.
6. Mukai, T., et al., *Scripta Mater.*, Vol. 39, 1249–1254, 1998.
7. Chapman, J. A., and D. V. Wilson, *J. Inst. Metals*, Vol. 91, 39–40, 1962-63.
8. Iwasaki, H., et al., *J. Japan Soc. Powder and Powder Metall.*, Vol. 11, 1350–1353, 1996.
9. Mabuchi, M., K. Kubota, and K. Higashi, *Mater. Trans, JIM*, Vol. 36, 1249–1254, 1995.
10. Nussbaum, G., et al., *Scripta Metall.*, Vol. 23, 1079–1084, 1989.

11. Mukai, T., et al., *Mater. Trans.*, Vol. 42, 1177–1181, 2001.
12. Armstrong, R., I. Codd, R. M. Douthwaite, and N. J. Petch, *Phil. Mag.*, Vol. 7, 45–58, 1962.
13. Hilpert, M., et al., *Magnesium Alloys and their Applications*, Werkstoff-informations gesellshaft, Hamburg, 319–322, 1998.
14. Furukawa, M., Z. Horita, M. Nemoto, and T. G. Langdon, *J. Mater. Sci.*, Vol. 36, 2835–2843, 2001.
15. Mabuchi, M., K. Ameyama, H. Iwasaki, and K. Higashi, *Acta Mater.*, Vol. 47, 2024–2057, 1999.
16. Mukai, T., M. Yamanoi, H. Watanabe, and K. Higashi, *Scripta Mater.*, Vol. 45, 89–94, 2001.
17. Mukai, T., H. Watanabe, and K. Hiagshi, *Mater Sci. Tech.*, Vol. 16, 1314–1319, 2000.

11

Enhancement of Crashworthiness in Cellular Structures

T. Miyoshi, M. Itoh, T. Mukai, S. Nakano, and K. Higashi

CONTENTS

Introduction

Recently, there has been great interest in using lightweight metallic foams (e.g., aluminium and magnesium) for automotive, railway, and aerospace applications where weight reduction and improved performance are needed.[1] Metallic foams also have potential for absorbing impact energy during vehicle crashes either against another vehicle or a pedestrian. To effectively absorb the impact energy, a material is required to exhibit an extended stress plateau. Thus, enhancement of the absorbed energy can be achieved by the extent of strain plateau and an increase in the plateau stress. The value of plateau stress can be increased by the selection of the matrix and its density. Thornton and Magee have demonstrated that the plateau stress of a closed-cell aluminum varied with composition and heat-treatment of the matrix.[2] Hagen and Bleck have also reported that the plateau stress could be increased with density using a commercially available closed-cell foam, ALPORAS.[3]

Gibson and Ashby[1] analyzed the relationship between the relative stress, σ_{p1}/σ_{y5}, and the relative density, ρ/ρ_s, assuming that plastic collapse occurs when the moment exerted by the compressive force exceeds the fully plastic moment of the cell edges, where σ_{pl} is the plastic-collapse stress, σ_{ys} is the yield stress of the cell wall (edge) material, ρ is the density of the cellular material, and ρ_s is the density of the cell wall (edge) material, respectively. The relationship between the relative stress and the relative density for open-celled material is given by Equation 11.1,

$$\frac{\sigma_{pl}}{\sigma_{ys}} = C\left(\frac{\rho}{\rho_s}\right)^{3/2} \tag{11.1}$$

where C is a constant. Gibson and Ashby[1] showed that the value of C is 0.3 from data of polyurethane foams and cellular metals. Gibson and Ashby also analyzed the relative stress for closed-cell material and reported the influence of membrane stress and cell fluid.[1] They suggested the membrane stress should be considered. Simone and Gibson analyzed the effects of solid distribution on the stiffness and strength of metallic foams[4] and also reported the effects of cell face curvature and corrugations.[5] The experimental data of closed-cell aluminum alloys[2] and a commercial closed-cell aluminum, ALPORAS, in the previous investigation[6] were shown to be in reasonable agreement with the values predicted for the case of $C = 0.3$.[1] The evidence indicates that the membrane stress has probably only a minor effect. On the other hand, Grenestedt has investigated the influence of imperfections such as wavy distortions of cell walls on the stiffness of closed-cell aluminum alloys.[7,8] Sanders and Gibson have also pointed out that the reduction in Young's modulus of aluminum foams is due to cell wall curvature and corrugation.[9] Therefore, it is noted that controlling the structure of cell walls is important for the enhancement of energy absorption. In this study, enhancement of crashworthiness in a closed-cell structure has been performed by an increase in the aspect ratio of cell-wall thickness against the cell-edge length with reduction of cell size. The crashworthiness in a modified foam is estimated by comparison with a conventional ALPORAS with the same relative density.

Structure of Modified Foam

The material used in the present study is a modified aluminum foam, ALPORAS (denoted as #M), which was produced by Shinko Wire Co. Ltd., Japan. The aluminum foam was manufactured by a batch-casting process. The chemical composition of the modified foam is Al-1.42,Ca-1.42,Ti-0.28,Fe (by mass%), which is the same as the regular ALPORAS. The details of fabrication of this foam have been reported elsewhere.[10] Typical structure of the foam is shown in Figure 11.1a. The average diameter of the cells was

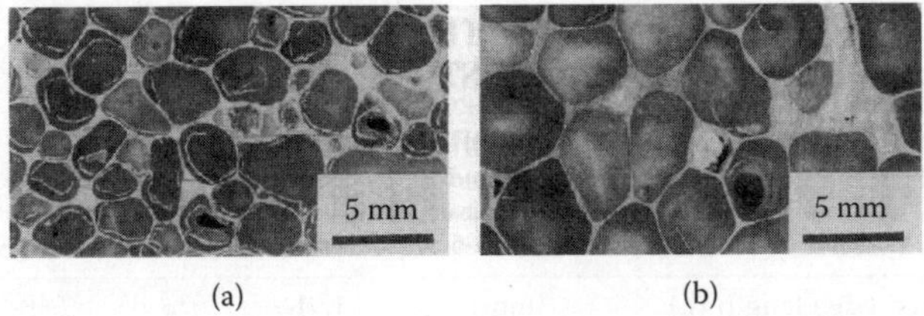

FIGURE 11.1
Typical cell structures of a modified foam (#M) (a), and the closed-cell aluminum foam (#C) (b) of a conventional ALPORAS.

measured to be ~3.0 mm using the method prescribed by the American Society for Testing Metals (ASTM) for the measurement of grain diameter in polycrystalline materials.[11] The relative density was about 0.106.

The referenced material is a conventional ALPORAS (denoted as #C), the structure of which is shown in Figure 11.1b. The relative density (0.105) of the foam is very close to the present foam, however, the average diameter (–4.5 mm) is larger than that of the present foam. In order to characterize the structures of two foams, optical microscopy has been used to measure the apparent edge length (denoted as L) and the thickness of cell walls from any 200 edges. A schematic illustration of this measurement is shown in Figure 11.2.

As shown in this figure, the apparent thickness was estimated for two points at ½ L and ¼ L. Note that the measured edge length of the modified foam is effectively reduced. The thickness of cell walls at ½ L and at ¼ L has also been evaluated. The thickness of the cell walls in the modified foam is slightly smaller than that in the conventional ALPORAS. Since the thickness of cell

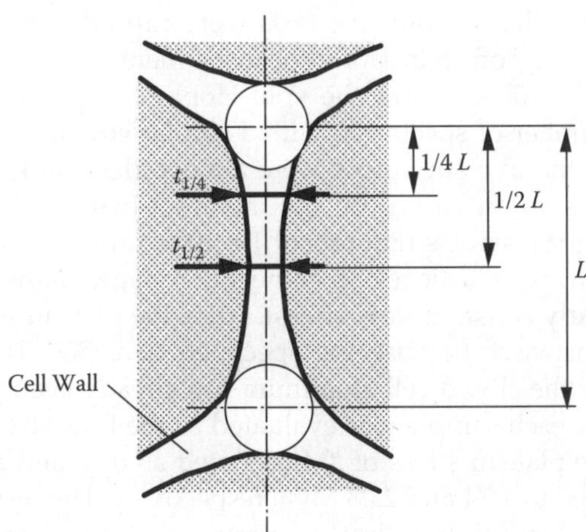

FIGURE 11.2
Schematic illustration of a cross-sectional structure in a closed-cell aluminum foam.

TABLE 11.1

The Measured Average Value of Edge Length and Wall Thickness
in a Modified Foam Comparing with a Conventional ALPORAS

		Modified Foam (#M)	Conventional ALPORAS (#C)
Measured edge length (L)	[mm]	1.91	2.68
Measured thickness of cell wall			
At 114 L ($t_{1/4}$)	[mm]	0.135	0.162
At 112 L ($t_{1/2}$)	[mm]	0.112	0.132
Aspect ratio	[mm]		
(t1l41L)		0.0693	0.0604
(t1l41L)		0.0585	0.0505

walls at ½ L is essentially greater than that at ¼ L for both foams, the cell walls have a thickness gradient. However, the modified foam exhibits a more uniform distribution of thickness compared with the conventional foam. The average values of the measured edge length and wall thickness were summarized in Table 11.1. The aspect ratio of the wall thickness against the edge length is also listed in Table 11.1. Note that the aspect ratio in the modified foam is higher than that in a conventional ALPORAS on the average.[12]

Compressive Behavior

Compressive tests were performed to evaluate the absorption energy. Specimens with dimensions of 16 × 11 mm were cut from each of the as-cast aluminum foams. The compressive tests were carried out at a quasi-static strain rate of 1×10^3 s^{-1} on an Instron-type instrument, and at a quasi-dynamic strain rate of 1×10^3 s^{-1} using the split-Hopkinson-pressure bar (SHPB) method for a number of specimens with a similar structure.

The curves of the five specimens were almost identical for both #M and #C. Typical stress–strain curves for the modified foam and a conventional ALPORAS at a quasi-static strain rate of 1×10^3 s^{-1} are shown in Figure 11.3. The diagram shows an elastic region at the initial stage, followed by a plateau region (with nearly constant flow stress). After the plateau region, the flow stress rapidly increases because the specimen densifies. The stress–strain characteristic for the closed-cell aluminum has already been reported.[13] The plateau stress for each sample was evaluated as the flow stress at a nominal strain of 0.2. The plateau stress of #M (denoted as σ_{Mp}) and #C (denoted as σcp) are measured to 1.74 and 2.03 MPa, respectively. The relative density of both samples is identical. It is clearly observed that the plateau stress in the modified foam exhibits a higher value than that in a conventional ALPORAS. The increase in the plateau stress seems to be due to the increase in the membrane stress of cell walls with an aspect ratio of $t_{1/2}/L$ (see Table 11.1).

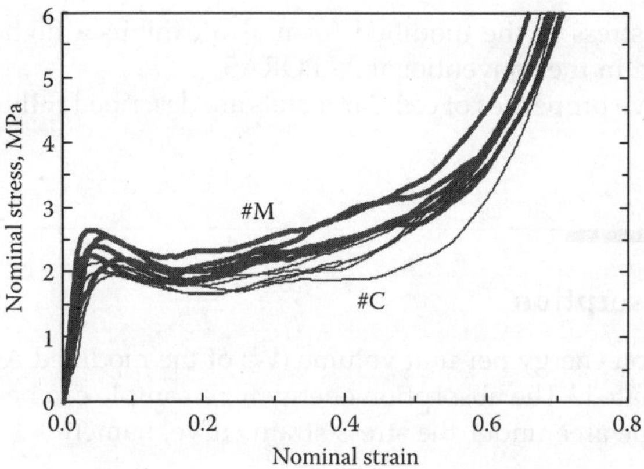

FIGURE 11.3
Nominal stress–strain curves at a quasi-static strain rate of 1×10^3 s^{-1}.

A number of stress–strain curves for both #M and #C at a dynamic strain rate of 1×10^3 s^{-1} are shown in Figure 11.4. In comparison to Figure 11.3, several different features can be observed:

1. the yield stress at the dynamic strain rate is higher than that at the quasi-static strain rate,

2. plateau strain (fp/) at the dynamic strain rate is slightly smaller than that at the quasi-static strain rate, and

3. gradual stress drop with strain can be seen in the curve at the dynamic strain rate.[6]

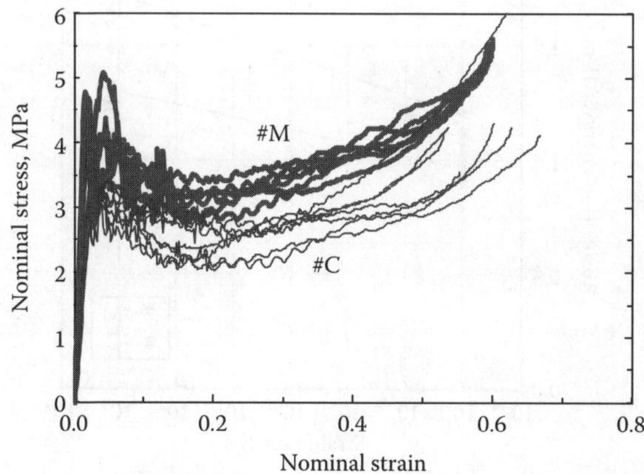

FIGURE 11.4
Nominal stress–strain curves at a dynamic strain rate of 1×10^3 s^{-1}.

The plateau stress in the modified foam also exhibits a higher value than that observed in the conventional ALPORAS.

Compressive properties of cellular metals are described fully in chapter 12.

Energy Absorption

The absorption energy per unit volume (W) of the modified ALPORAS was further evaluated.[1] The absorption energy for a sample can be calculated by integrating the area under the stress-strain curve, namely

$$W = \int_0^\varepsilon \sigma(\varepsilon) \cdot d\varepsilon \qquad (11.2)$$

The average values of absorption energy per unit volume of ALPORAS at a strain of 0.5, and the dynamic strain rate for #C and #M are calculated as 1.30 and 1.72 MJ/m³, as shown in Figure 11.5. The absorption energy at quasi-static strain rate of the conventional foam is very close to the value formerly reported by Hagen and Bleck.[3] The value of W in the modified foam at the dynamic strain rate, #M, is about 32% higher than that in #C. Enhancement of energy absorption can be achieved using the present modified structure. The selection of cellular materials for applications, such as cycle helmet inner liners,[1] bumpers for automobiles, or motor cycles, are

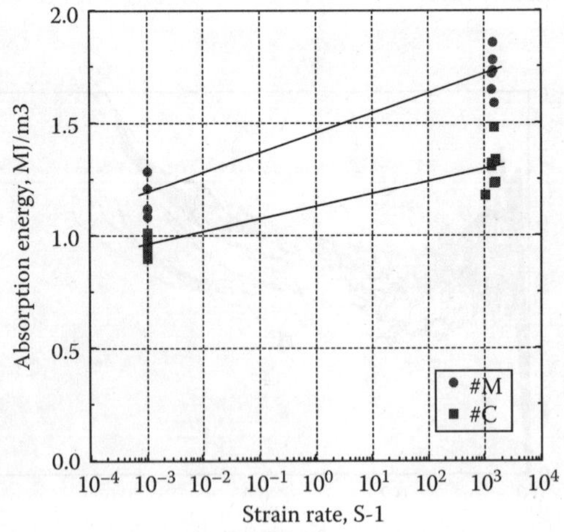

FIGURE 11.5

Absorption energy per unit volume of ALPORAS at a strain of 50%.

based on energy absorption characteristics. The results previously reported for ALPORAS suggested that there was a large difference in mechanical strength and absorption energy obtained at a quasi-static and a dynamic strain rate.[6] The examination at dynamic strain rate of the present foam is underway. Optimization of cellular structures is important for the enhancement of energy absorption in a closed-cell metallic foam.

Summary

Metallic foams have potential for absorbing impact energy. To effectively absorb the impact energy, a material must exhibit an extended stress plateau. Thus, enhancement of the absorbed energy can be achieved by extension of the strain plateau, and increasing the plateau stress. In this study, modification of the structure in a commercially available closed-cell aluminum foam, ALPORAS, was performed. The edge length of this modified foam is obviously reduced from that of a conventional ALPORAS, while the aspect ratio of the wall thickness to the edge length increases. From the compressive tests at a quasi-static rate and a dynamic strain rate, the plateau stress of the modified foam was found to exhibit a marked increase compared to that of the conventional foam. The absorption energy per unit volume (W) of the modified foam is 40% higher than that of a conventional foam.

References

1. Gibson, L. J., and M. P. Ashby, in *Cellular Solids, Structure and Properties—Second edition*, Cambridge University Press, Cambridge, UK, 1997.
2. Thornton, P. H., and C. L. Magee, *Metal!. Trans. A.*, Vol. 6A, 1253, 1975.
3. Hagen, R. V., and W. Bleck, *Mat. Res. Soc. Symp. Proc.*, ed. D. S. Schwartz, D. S. Shih, A. G. Evans, and H. N .G. Wadley, Materials Research Society, Vol. 521, 59, 1998.
4. Simone, A. E., and L. J. Gibson, *Acta Mater.*, Vol. 46, 2139, 1998.
5. Simone, A. E., and L. J. Gibson, *Acta Mater.*, Vol. 46, 3929, 1998.
6. Mukai, T., H. Kanahashi, T. Miyoshi, M. Mabuchi, T. G. Nieh, and K. Higashi, *Scripta Mater.*, Vol. 40, 921, 1999.
7. Grenestedt, J. L., *J. Mech. Phys. Solids*, Vol. 46, 29, 1998.
8. Grenestedt, J. L., *Mat. Res. Soc. Symp. Proc.*, ed. D. S. Schwartz, D. S. Shih, A. G. Evans, and H. N. G. Wadley, Materials Research Society, Vol. 521, 3, 1998.
9. Sanders, W., and L. Gibson, *Mat. Res. Soc. Symp. Proc.*, ed. D. S. Schwartz, D. S. Shih, A. G. Evans, and H. N. G. Wadley, Materials Research Society, Vol. 521, 53, 1998.

10. Miyoshi, T., M. Itoh, S. Akiyama, and A. Kitahara, *Mat. Res. Soc. Symp. Proc.*, ed. D. S. Schwartz, D. S. Shih, A. G. Evans, and H. N. G. Wadley, Materials Research Society, Vol. 521, 133, 1998.
11. ASTM Designation E 112-82.
12. Miyoshi, T., M. Itoh, T. Mukai, R. Kanahashi, H. Kohzu, S. Tanabe, and K. Higashi, *Scripta Mater.*, Vol. 41, 1055, 1999.
13. Sugimura, Y., J. Meyer, M. Y. He, H. B. Smith, J. L. Grenstedt, and A. G. Evans, *Acta Mater.*, Vol. 45, 5245, 1997.

12

Compressive Properties of Cellular Metals

Mamoru Mabuchi

CONTENTS

Introduction

Recently, there has been a considerable increase in interest in cellular metals. Cellular metals are super-light metals exhibiting unique properties, such as high energy absorption.[1] Applications of cellular metals are wide ranging, e.g., impact energy absorbers, silencers, flame arresters, heaters, heat exchangers, constructional materials, etc.

Typical cellular metals are shown in Figure 12.1. Cellular solids are divided into two types; one is a closed-cellular solid where each cell is sealed off from its neighbor (Figure 12.1a), the other is an open-cellular solid whose cells connect through open faces (Figure 12.1b).

To date, the mechanical properties of cellular metals have been extensively investigated.[1] In general, a cellular metal shows an elastic region in the initial stage of applied strain, and then a plateau region with a nearly constant flow stress (collapse stress), and finally, a densification region where the flow stress significantly increases when the cellular metal is compressed. The collapse stress in the plateau region is strongly affected by the density of the material. Gibson and Ashby[1] analyzed the collapse stress of an open-cellular metal using bending of struts, showing that the collapse stress of an open-cellular metal is proportional to the 3/2 power of the density. A closed-cellular metal also exhibits almost the same behavior as an open-cellular metal. However, the

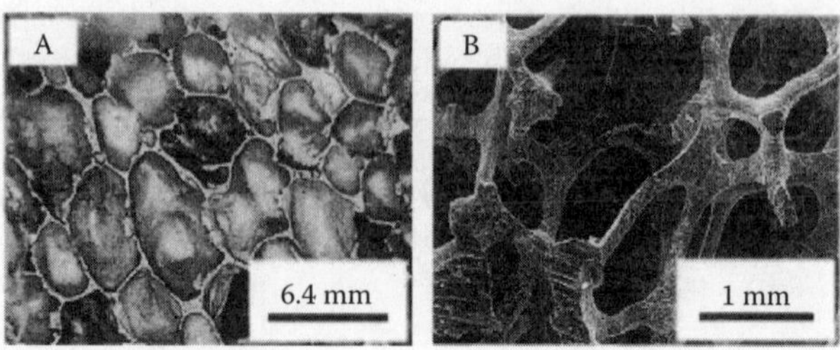

FIGURE 12.1
Cellular metals, (A) closed-cell and (B) open-cell.

collapse stress in the plateau region is considered to be affected not only by the density, but also by the solid material and the cell structure. In this chapter, effects of the solid materials and the cell structure on compressive properties of a cellular metal are investigated.

Effect of the Solid Material

Compressive properties of the cellular SG91A aluminum alloy and cellular AZ91 magnesium alloy with the open-cell structure were investigated,[2] where both cellular alloys had the same cell structure. The nominal stress-nominal strain curves are shown in Figure 12.2 for the cellular aluminium alloy, and in Figure 12.3 for the cellular magnesium alloy—five specimens of each material

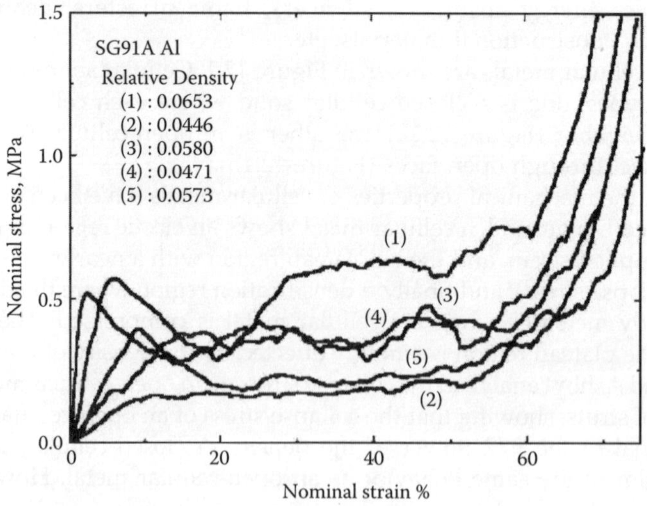

FIGURE 12.2
The nominal stress–nominal strain curves for the cellular SG91A Al alloy.

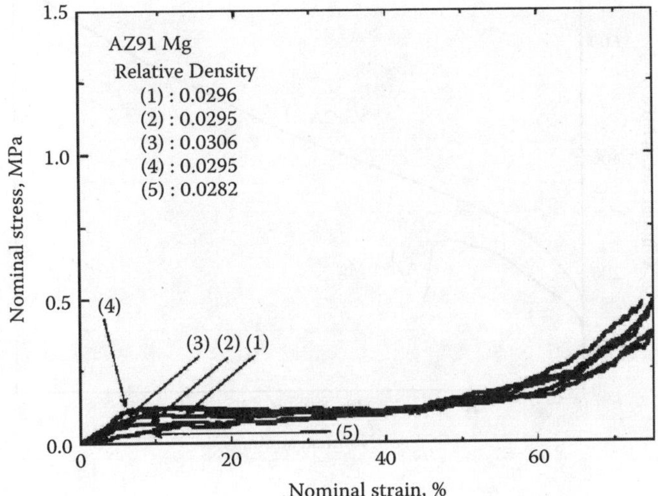

FIGURE 12.3
The nominal stress–nominal strain curves for the cellular AZ91 Mg alloy.

were measured. The cellular alloys display an elastic region at the initial stage, and then a large plateau region with a nearly constant flow stress to a large strain of about 60%. The yield stresses were 0.17 ~ 0.58 MPa for the cellular aluminium alloy and 0.07 ~ 0.13 MPa for the cellular magnesium alloy. The cellular aluminium alloy showed higher yield stress and plateau stress than the cellular magnesium alloy. After the plateau region, the flow stress rapidly increased because of densification for both cellular alloys. This trend of the open-cellular alloys is similar to those of other cellular metals.

The nominal stress–nominal strain curves of the solid SG91A aluminium and AZ91 magnesium alloys with a relative density of 100% are shown in Figure 12.4. The yield stresses of the SG91A aluminium and AZ91 magnesium solids were 150 and 120 MPa, respectively. The AZ91 magnesium solid fractured at $\varepsilon = 14\%$. However, the SG91A aluminium solid did not fracture until $\varepsilon = 50\%$. The compressive test was stopped at $\varepsilon = 50\%$. Although the AZ91 magnesium solid showed much lower ductility than the SG91A aluminium solid, the strain to densification of the cellular AZ91 magnesium was almost the same as that of the cellular SG91A aluminium.

The cellular aluminium alloy exhibited larger flow stress in a plateau region than the cellular magnesium. However, the difference in yield stress between the aluminium alloy solid and the magnesium alloy solid was little. The relationship between the relative stress, $\sigma_{pl}/\sigma_{ys'}$ and the relative density, $\rho/\rho_{s'}$ is given by Equation 12.1

$$\frac{\sigma_{pl}}{\sigma_{ys'}} = C\left(\frac{\rho}{\rho_{s'}}\right)^{3/2}$$

(12.1)

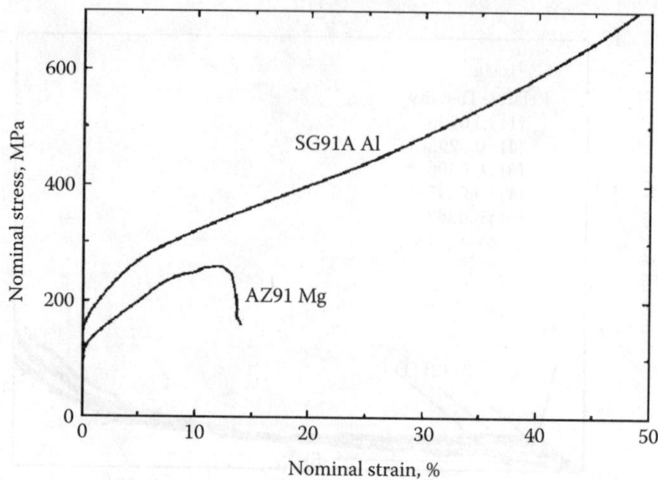

FIGURE 12.4
The nominal stress–nominal strain curves of the solid SG91A Al and AZ91 Mg alloys with the relative density of 100%.

where σ_{pl} is the collapse stress of a cellular metal, σ_{ys} is the yield stress of a solid, ρ is the density of a cellular metal, ρ_s is the density of a solid, and C is a constant. This equation indicates that the flow stress in the plateau region of cellular metals is proportional to the three-seconds power of the relative density. Therefore, it is suggested that the larger flow stress in a plateau region for the cellular Al alloy is mainly attributed to the larger relative density.

The normalized stress–nominal strain curves of the cellular aluminium alloy and cellular magnesium alloy are shown in Figure 12.5, where the normalized stress is the nominal stress divided by the yield stress of the solid and the three-seconds power of the relative density. It can be seen that the normalized stress–nominal strain curves of the cellular magnesium alloy are in agreement with those of the cellular aluminium alloy. The stress–strain relation of the magnesium alloy solid was different from that of the aluminium alloy solid; in particular, the magnesium alloy solid showed much lower ductility than the aluminium alloy solid. However, the stress–strain relation of the cellular magnesium alloy was the same as that of the cellular aluminium alloy by compensation with the yield stress of the solid and the relative density. Therefore, it is concluded that the mechanical properties of the cellular metals are not affected by ductility of the solid. This suggests that once the cell edge collapses at the yield point of the solid, the collapsed edge has little ability to bear the load, and bends easily by a low stress; as a result, the mechanical properties of cellular metals are independent of ductility and strain-hardening behavior of the solid.

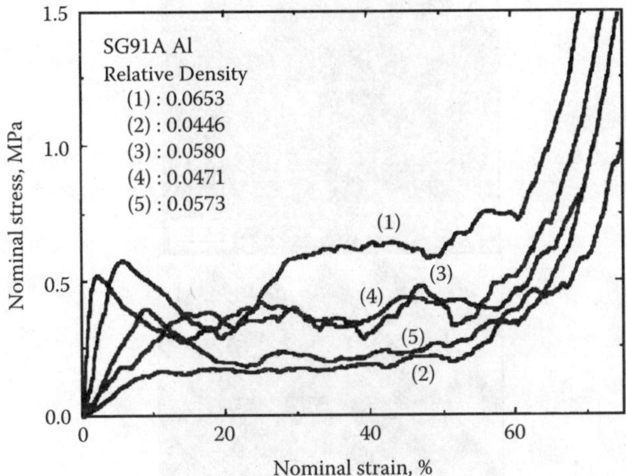

FIGURE 12.5
The normalized stress–nominal strain curves of the cellular SG91A Al alloy and cellular AZ91
Mg alloy.

Effect of the Cell Structure

Three kinds of open-cellular magnesium alloys—type A is a cellular magnesium alloy with random cell structure; type B is a cellular magnesium alloy with controlled cell structure for which an angle between the struts and the load direction is 45°; and type C is a cellular magnesium alloy with controlled cell structure for which an angle between the struts and the load direction is 0° (90°)—were fabricated by casting[3] (see Figure 12.6).

The nominal stress–nominal strain curves of the cellular magnesium alloys by compressive tests are shown in Figure 12.7. The type B structure showed almost the same stress–strain curve as the type A structure, indicating that the dominant deformation mode of a cellular magnesium alloy with random cell structure is the same as that of a cellular magnesium alloy with controlled cell structure for which an angle between the struts and the load direction is 45°. The collapse stresses were about 0.1 MPa for the type A and the type B structures. For the type C structure, however, the flow stress decreased after a sharp peak, and then it rapidly increased with increasing strain. This fluctuation of the flow stress is repeated in the plateau region. This trend is the same as that in the cellular epoxy.[4] The collapse stress (first peak stress) was 0.4 MPa for the type C structure. The collapse stress of the type C structure was 4 times higher than those for the type A and the type B structures. Recently, Markaki and Clyne[5] revealed that the mechanical properties of a cellular aluminium alloy were strongly affected by the microstructure. In the present investigation, however, the density and microstructure

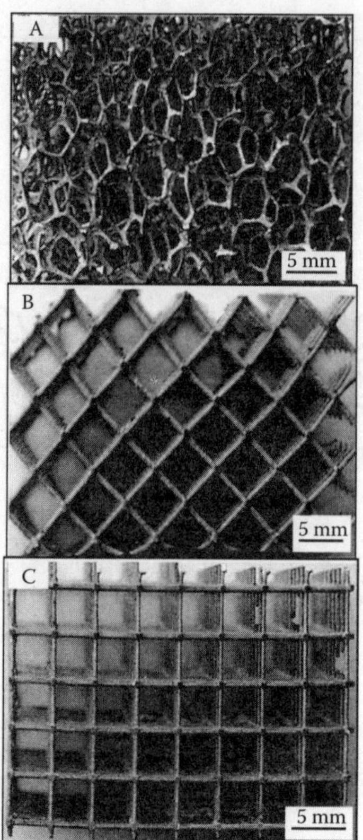

FIGURE 12.6
Three kinds of open-cellular Mg alloys.

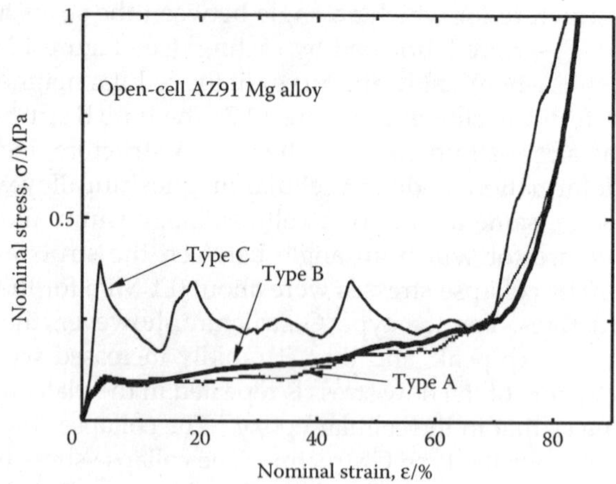

FIGURE 12.7
The nominal stress–nominal strain curves of the cellular Mg.

of the cellular magnesium alloys were the same. Clearly, the cell structure strongly affects the collapse stress in the plateau region of cellular metals.

Another important result shown in Figure 12.7 is that the densification strain, which is the strain to the onset of the densification region, is almost the same for three kinds of cellular magnesium alloys. That is to say that the densification strain is not affected by the cell structure. It is, therefore, suggested that the densification strain mainly depends on the relative density of the material.

Summary

The stress–strain relation of cellular magnesium alloy was almost the same as that of the cellular Al alloy by compensation with the yield stress of the solid and the relative density. Therefore, it is concluded that the mechanical properties of the cellular metals are not affected by ductility of the solid. The cell structure strongly affected the collapse stress in the plateau region of cellular metals. However, the densification strain was not affected by the cell structure.

References

1. Gibson, L. J., and M. F. Ashby, *Cellular Solids, Structure and Properties*, Cambridge University Press, Cambridge, 175, 1997.
2. Yamada, Y., et al., *Mater. Sci. Eng. A*, A272, 455, 1999.
3. Yamada, Y., et al., *Mater. Trans.*, in press, 2002.
4. Yamada, Y., et al., *Philos. Mag. Lett.*, Vol. 80, 215, 2000.
5. Markaki, A. E., and T. W. Clyne, *Acta Mater.*, Vol. 49, 1677, 2001.

13

Heavily Deformable Al Alloy

Osamu Umezawa

CONTENTS

A Recyclable Design for Sustainable Development

Problems in Scrap Metal Recycling

Products generally consist of many kinds of materials. However, it is necessary to save resources and energy, and so there is a general move toward the use of recycled materials. Engineering challenges for the ideal recycling system are to reduce impurity content from scrap melt, to immunize metals against impurities or make metals innocuous to impurities, and finally, to replace conventional metals with inherently recyclable metals. As a design for *ecomaterials* (environment conscious materials)[1] recyclable design for sustainable development has been proposed, where the materials and products are in harmony with the natural environment and provide a simplified recycling route. The concept has been discussed from the point of view of design factors, i.e., recyclability and ecology for raw materials, and simplification of recycling and scrapping, balanced against the desired properties.[2] Aluminum products, as well as steels, are some of the early metals that have taken up the challenge of recycling. In the case of aluminum, recycling is desirable, since producing aluminum from virgin mineral sources, such as bauxite, consumes large amounts of energy. However, the incorporation of secondary (scrap) aluminum has been almost limited to cast materials, and dilution with raw material has been inevitable. In the 1990s, about 30% by weight of aluminum products were produced from scrap metals in Japan. The use of scrap metal will increase, so secondary metal products should be manufactured not only using cast materials, but also, wrought ones.

Upgrade Recycle Design for Al-Si-X Alloys

In the case of aluminum, silicon, iron, and copper, are major detrimental elements for recycling aluminum products. Silicon and iron presents problems in aluminum alloy recycling due to the poor workability of aluminum incorporating these elements, and the general difficulty in removing them from stock materials. Al-Si-X alloys are one of the major cast material systems, and include almost no undesirable elements for the classification *ecomaterial*. However, removal of Si and Fe from scrapped aluminum products is difficult and costly. The Al-Si system has multiple phases with low mutual solid solubility, and are effectively an in-situ metal-metal composite, making it an ideal model for recycling whilst maintaining control of the balance of properties, such as strength and elongation, without detrimental elements for the *ecomaterials;* fine microstructure with plural phases, *mesocomplex structure*, is one of the candidates for alloy design.[2] Hence, Al-Si based alloys have great advantage in recyclable material design. However, heavy cold-working cannot be used, since this would cause severe cracking in the primary Si crystals and coarse intermetallic compounds.

Fine Microstructure Development

To improve the mechanical properties of Al-Si-X cast materials, microstructural modifications have been commonly achieved by the addition of elements such as strontium and phosphorus into the melt, or by hot-forging and subsequent long solution heat-treatment, as shown in Figure 13.1. However, the ductility of these treated materials is not enough, and it is difficult to apply these treatments to wrought materials. Coarse silicon crystals and/or acicular Al-Si-Fe intermetallic compounds cause poor ductility and often give rise to fatigue crack initiation sites, as well as inclusions. To improve the workability and mechanical properties, the silicon crystals and compounds must be refined to avoid sample fracture due to their cracking.

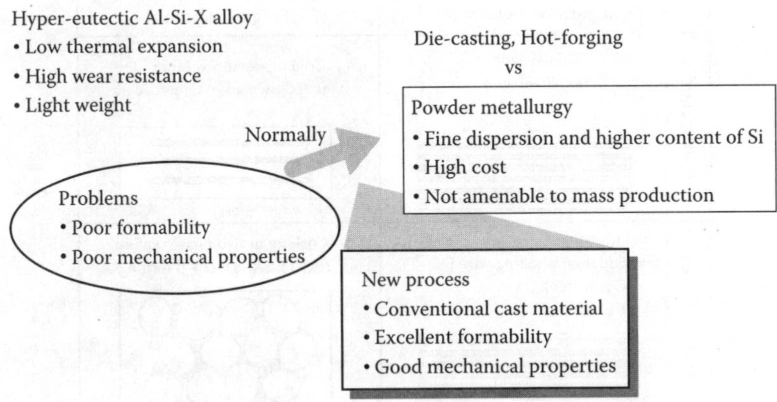

FIGURE 13.1
Development of Al-Si-X forgings from cast material.

The most promising method of refining microstructures is by rapid solidification. Spray-forming and thixoforming have been used to refine the silicon in Al-Si alloys. However, rapid solidification methods are not necessarily amenable to mass production, so the present work deals with a novel thermomechanical treatment for hyper-eutectic Al-Si-X cast alloys, which could be mass produced. Umezawa and Nagai[3-4] have proposed repeated thermomechanical treatment (RTMT) to produce a heavily deformable hyper-eutectic Al-Si material, as shown in Figure 13.2. Repeated thermomechanical treatment also provides a fine dispersion of Al_5SiFe compounds in hyper-eutectic Al-Si-(Fe, Cu) cast alloys.[4-6] Not only Si, but also iron and copper are major impurities in these secondary aluminum cast alloys. In this chapter, the influence of microstructural refinement by the repeated thermomechanical treatment on both tensile and high-cycle fatigue properties for these alloys is reviewed.

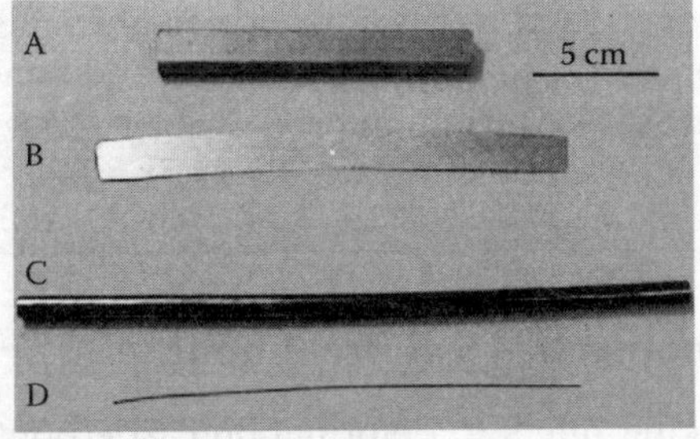

FIGURE 13.2
Al-12.6 mass%Si repeated thermomechanical treatment materials: annealed plate (a) and its cold-rolled sheet (b), and annealed rod (c) and its cold-swaged wire (d).

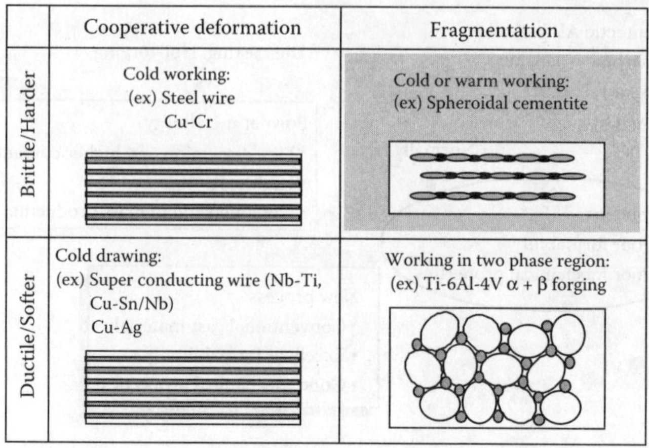

FIGURE 13.3
Second-phase structure developed by plastic workings.

Concept of Mesocomplex Structure[2]

Mechanical properties can be generally controlled by changing the microstructure of any particular material. Design and control of the *mesocomplex structure*, where the secondary phases are distributed in a matrix with high density and fine structure, are illustrated in Figure 13.3. Heavy working plays an important role on building the fine microstructure.

1. Microscopic distribution of secondary phase: When the secondary phase is deformable, the increase of plastic strain leads to the decrease of lamella spacing. When the secondary phase cannot be deformed, the plastic strain induces cracking of this phase.

2. Deformation of matrix and a role of secondary phase as a pinning site: The rapid increase of plastic strain in the matrix makes a dynamic recovery and recrystallization in fine scale grain size. The secondary phase is a site for local deformation and a pinning site for the dynamic recovery in matrix.

Repeated Thermomechanical Treatment

Generally, hot-working refers to deformation carried out under conditions of temperature and strain rate. Since recovery processes occur substantially during the deformation process, large strains can be achieved with essentially no strain hardening. This results in a decrease in the energy required

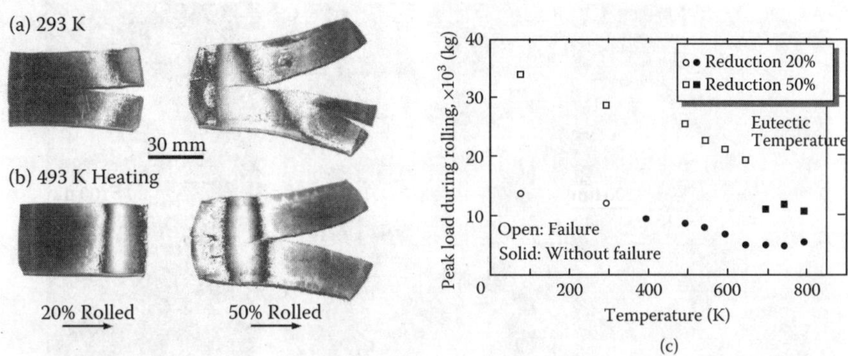

FIGURE 13.4

One-pass rolling of Al-12 mass%Si cast material; test pieces at (a) 293 K and (b) 493 K, and (c) deformation resistance at various temperatures.

to deform the metal and an increased ability to flow without cracking. Thus, the higher the working temperature is, the less cracked Si crystals are detected. Severe cracking in silicon crystals occurs under cold-working operations, as shown in Figure 13.4. The cracks may lead to sample fracture, but could be useful in refining the Si crystals. In order to avoid fracture, the cold-working operations must be carried out with intermediate annealing steps.

Therefore, cold-working operations were carried out in several steps with intermediate annealing operations introduced to soften the cold-worked materials. This sequence of repeated cold-working and annealing is called repeated thermomechanical treatment. Cold-working operations by flat-rolling, grooved-rolling, or swaging are carried out in a number of multiple passes or steps at, or below room temperature, forming specimens into either rods or plates. Annealing was at 793 K for 3.6 ks followed by water quenching (WQ). Each cold reduction was less than 20% in section area, and the cold-work-anneal cycle was repeated over six times. The total reduction in section area was about 80%.

The test materials were various kinds of hyper-eutectic Al-Si-(Fe, Cu) castings. The chemical compositions of the casts are listed in Table 13.1. Figure 13.5 shows their microstructures. Each cast material contains coarse

TABLE 13.1

The Chemical Compositions of Test Materials

	Concentrations (Mass%)							
Materials	Si	Fe	Cu	Mg	Mn	Cr	Zn	Ti
S1	13.3	1.85	0.028	0.004	0.008	0.006	<0.001	0.005
S2	11.9	0.15	4.31	<0.001	0.003	0.005	<0.001	0.006
S3	19.7	0.18	4.71	0.003	0.006	0.006	0.001	0.006
L1	12.5	2.14	0.004	0.010	<0.001	<0.001	0.004	0.011
L2	12.0	0.11	3.43	0.009	<0.001	<0.001	0.003	0.010
L3	20.6	0.13	0.027	0.008	0.006	0.006	0.005	0.005

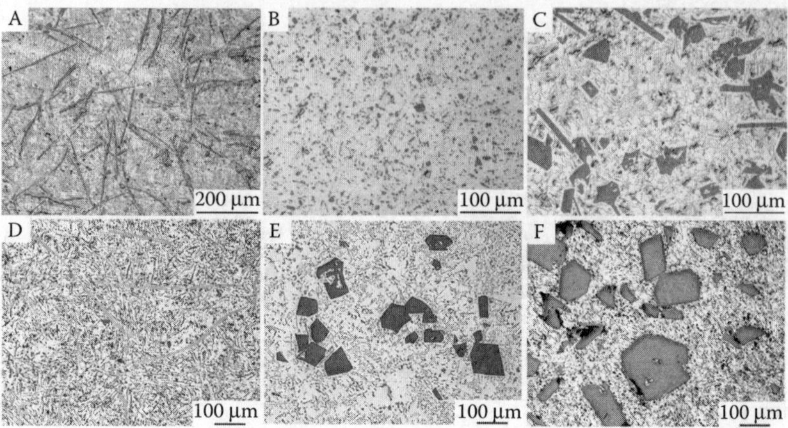

FIGURE 13.5
Microstructures of cast materials: (a) S1, (b) S2, (c) S3, (d) L1, (e) L2, and (f) L3.

primary Si crystals and/or primary compounds. The primary Si forms hexagonal-like crystals between 20 μm and 200 μm in diameter. The eutectic Si phase exhibits a needle or plate-like structure. The intermetallic compounds, β phase (Al_5SiFe), are acicular in form a few hundred μm in length.

In the case of large ingots cooled slowly, cast materials exhibit extremely low workability at, or below, room temperature, since they involve very large silicon crystals and compounds. A hot-working step needs to be introduced to provide a certain cold-workability for repeated thermomechanical treatment to produce a 20% section reduction without visible cracking. The casts are heated at a temperature between 673 K and 693 K, and are extruded with a working strain of 0.97 ($\eta = \ln A_0/A$, A_0: section area of sample; A: section area of worked sample). The extruded rods are annealed at 793 K for 3.6 ks, and then the repeated thermomechanical treatment is conducted.

Fragmentation of Silicon and Compounds

During cold-working, each primary silicon crystal cracks into a few pieces, decreasing in size. The eutectic silicon and associated compounds are also broken. The void formed by cracking heals up after repeated annealing and cold-working. The cracks in the primary silicon crystal tend to occur normal to the rolling direction.

In repeated thermomechanical treatment materials, primary silicon crystals are refined to less than a few tens of microns in diameter, and eutectic silicon crystals are broken and dispersed as fine particles (see Figure 13.6). Intermetallic compounds are also divided into the pieces of less than a few tens of microns in length. Most of the silicon crystals and compounds are a few microns in size, and aligned along the rolling direction (RD). The secondary phases are thus

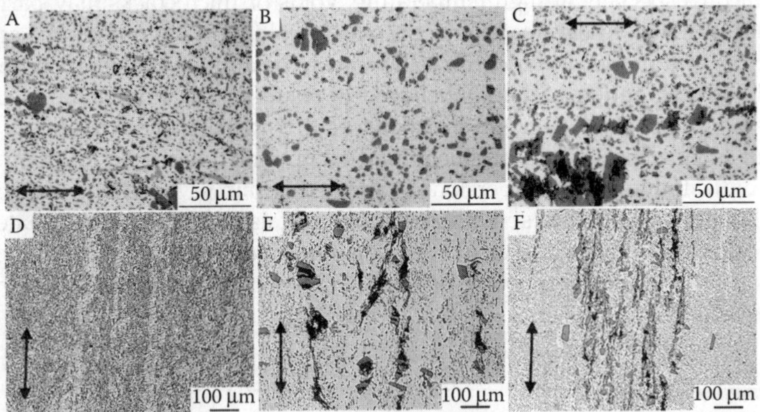

FIGURE 13.6
Microstructures of repeated thermomechanical treatment materials in the longitudinal section:
(a) 13Si-2Fe, (b) 12Si-4Cu, (c) 20Si-4Cu, (d) 13Si-2Fe: L1, (e) 12Si-4Cu: L2, and (f) 20Si: L3.
Photographs d–f show the casts cooled slowly. Arrows show the longitudinal direction.

distributed in the matrix with high density and fine size. In addition, they are spheroidized. In slowly cooled casts, however, the fragmentation and dispersion of primary Si crystals and/or compounds in the aluminum matrix through repeated thermomechanical treatment are insufficient for most applications (Figure 13.6d–f). The cracks in the primary Si remain, and are not fully filled by the aluminum matrix during repeated thermomechanical treatment.

When the secondary silicon phase in the soft Al matrix is non-deforming, the plastic work causes cracking in this secondary phase. The succeeding heat treatment softens the aluminum matrix again, as illustrated in Figure 13.7. Since the liner thermal expansion coefficient of aluminum is much higher than that

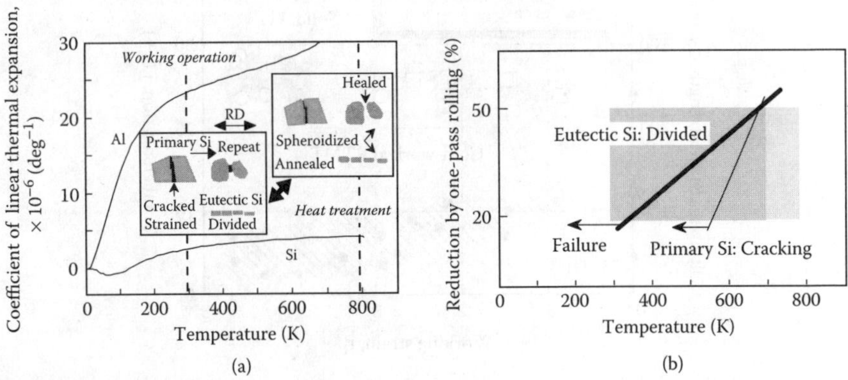

FIGURE 13.7
Schematic illustration of silicon cracking and Al matrix annealing: (a) thermal expansion coefficient of aluminium and silicon, and (b) silicon cracking temperature.

of silicon, especially at high temperature, voids or microcracks may partially heal during these heat treatments. Successive cold working presses the cracks together and the separate segments flow with the aluminum. Hence, repeated cold working and annealing is believed to result in the annihilation of these cracks. Consequently, several iterations of thermomechanical treatment can result in fine dispersions of silicon and compound phases. In addition, the speroidization of these dispersions is attained within several hours, although ten hours are needed to speroidize silicon crystals for the Al-Si cast material. Repeated thermomechanical treatment is effective for not only breaking up silicon crystals and compounds, but also healing cracks and speroidizing them.

Plasticity and Tensile Properties

Al-Si-X alloys with refined microstructure exhibit good workability, tensile strength, and elongation (see Figure 13.8). In both plate and rod forms, great plasticity, over 90% reduction in section area, is available. There is no severe cracking in the Si crystals in a cold-worked repeated thermomechanical treatment material (see Figure 13.9), but many voids are observed in the aluminium matrix.

Tensile testing was conducted in a motor-driven testing machine at 293 K (in air) and 77 K (immersed in liquid nitrogen) under displacement control. The displacement rate for each specimen was chosen to correspond to an engineering strain rate of approximately 4×10^{-4} per second within the plastic regime. Tensile specimens were machined parallel to the longitudinal direction.

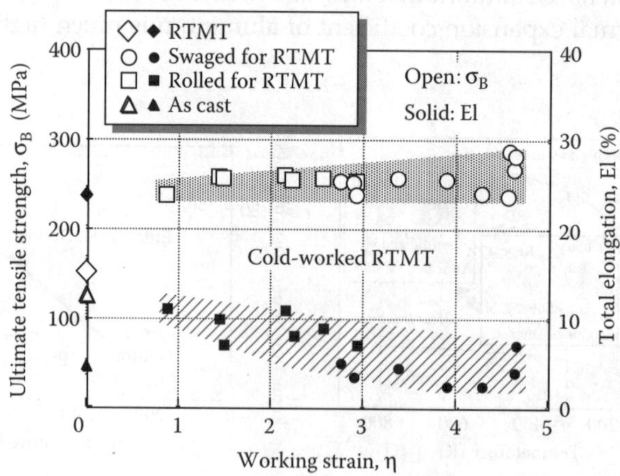

FIGURE 13.8

Tensile strength and elongation for repeated thermomechanical treatment and its cold-worked samples of Al-12.6 mass%Si alloy.

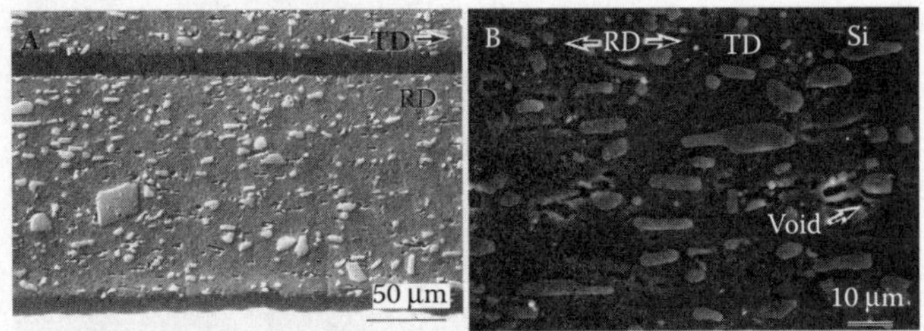

FIGURE 13.9
SEM micrographs of a cold-rolled repeated thermomechanical treatment sample ($\eta = 3$): (a) back scatter image for the RD section, and (b) secondary electron image for the TD section.

All cast materials showed poor elongation and early fracture caused by cracking of the secondary phase. Although the ultimate tensile strength of the repeated thermomechanical treatment material was almost the same as that of the cast one, the elongation of the repeated thermomechanical treatment material was much higher than that of the cast one, as shown in Figure 13.10. Local elongation and uniform elongation increase with repeated thermomechanical treatment. At lower temperatures, especially, microstructural refinement could result in the avoidance of early fracture as observed for the cast material. An increase in the flow stress with decreasing temperature results in a greater strain to failure for the repeated thermomechanical treatment samples. In general, increases in the strain hardening rate lead to an increase in the strain to failure. A geometrical instability occurs in a tensile test when the strain hardening rate equals the true stress. The repeated thermomechanical treatment samples show the necking instability, but the cast samples fractured before reaching it (see Figure 13.11). Therefore, the repeated thermomechanical treatment materials have a significantly enhanced ductility and strength, since early fracture is overcome in the

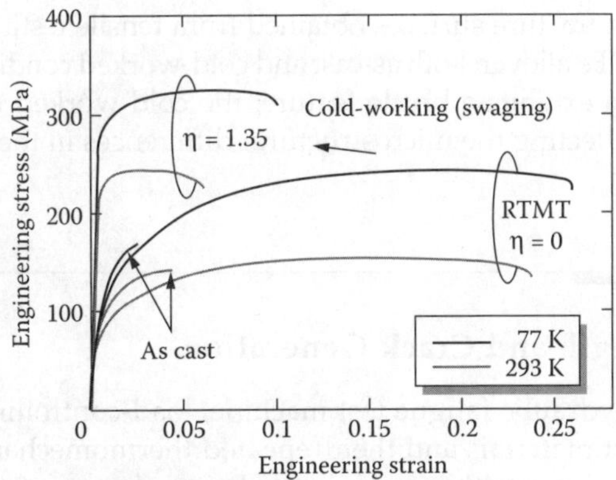

FIGURE 13.10
Stress–strain curves of Al-12.6 mass%Si alloy.

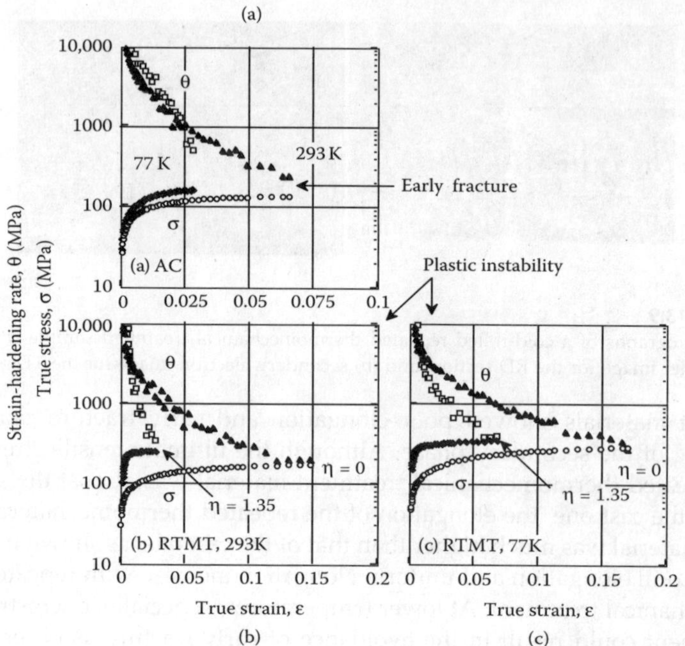

FIGURE 13.11
True stress–true plastic strain curves and strain-hardening rate curves from data in Figure 13.10: (a) as cast, (b) repeated thermomechanical treatment samples at 293 K, and (c) repeated thermomechanical treatment samples at 77 K.

repeated thermomechanical treatment condition. Microstructural modification by repeated thermomechanical treatment is effective in improving tensile properties.

Figure 13.12 shows elemental maps obtained by electron probe microanalysis, and the fracture surfaces obtained from tensile testing Al-7 mass %, Si-1 mass %, and Fe alloy in both as-cast and cold-worked conditions. Although the cast material exhibits a brittle feature, the cold-worked material shows a ductile one, reflecting the microstructural differences in the two materials.

Fatigue Strength and Crack Generation

Using a servo hydraulic fatigue test machine, load-controlling fatigue tests for the large cast materials and their repeated thermomechanical treatment ones were carried out with a stress ratio, R ($\sigma_{min}/\sigma_{max}$) = 0.01, supplied as a sine wave at, and below, 296 K. No large difference in fatigue strength was observed in either the low cycle or high cycle regime (see Figure 13.13).

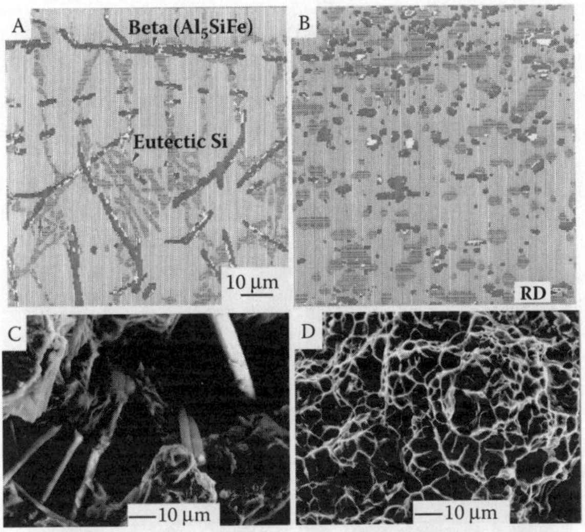

FIGURE 13.12

EPMA maps (a, b) and fracture surfaces from tensile testing (c, d) for Al-7Si-1Fe with as-cast (a, c) and cold-worked ($\eta = 5.8$) repeated thermomechanical treatment (b, d) materials.

The ratio of 10^7 cycles fatigue strength to ultimate tensile strength is about 0.5. The lower the temperature, the better the balance between tensile and fatigue properties.

Cracking and de-cohesion of the remaining coarse silicon crystals or aggregates of compounds give rise to fatigue crack initiation (see Figure 13.14). To improve the fatigue strength of the repeated thermomechanical treatment materials, further refinement of these large silicon crystals and/or compounds is required.

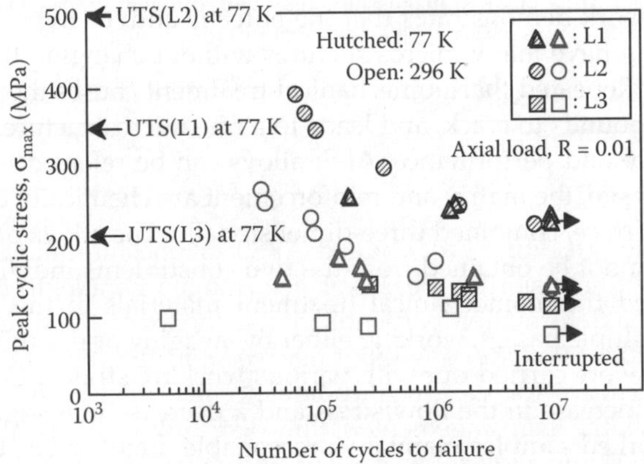

FIGURE 13.13

S-N data of repeated thermomechanical treatment materials at 77 K and 296 K.

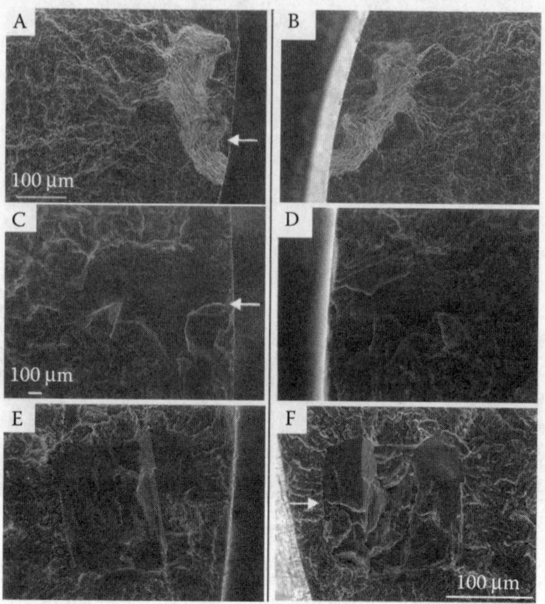

FIGURE 13.14

Matching halves of near fatigue crack initiation sites at 77 K: (a), (b) L1, σ_{max} = 249 MPa, Nf = 153,580 cycles, (c), (d) L2, σ_{max} = 393 MPa, Nf = 61,820 cycles, and (e), (f) L3, σ_{max} = 133 MPa, Nf = 1,604,240 cycles. Arrows indicate the crack initiation site.

Technological Applications

The present work demonstrates that the eutectic and/or hyper-eutectic Al-Si-X alloys can have many microstructures without changing the elemental composition. Repeated thermomechanical treatment causes the silicon crystals and compounds to crack, and leads to a fine microstructure. In terms of microstructure and performance, Al-Si alloys can be referred to as a metal matrix composite; the matrix and reinforcement are chemically distinct with a definite interface, combined three-dimensionally. The alloys possess properties that can not be obtained from the two constituents individually.

For repeated thermomechanical treatment materials in the as-annealed condition, multiple passes, working either by swaging or flat-rolling at room temperature, were carried out with various working strains. Cold-working results in an increase in the flow stress and a decrease in elongation. However, cold-worked samples maintain a reasonable ductility, i.e., both several percent uniform and local elongation. The work hardening of Al matrix may be responsible for the higher strength in the cold-worked materials. Figure 13.15 shows transmission electron miloscope images of a heavily worked material, and Figure 13.16 shows the electron backscatter diffraction pattern (EBSP)

FIGURE 13.15
TEM bright field images of cold-swaged ($\eta = 2.3$) repeated thermomechanical treatment S1 material (a, b) and its aging at 423 K for 86.4 ks (c, d) in the transverse section. Arrows indicate Al_5SiFe compound.

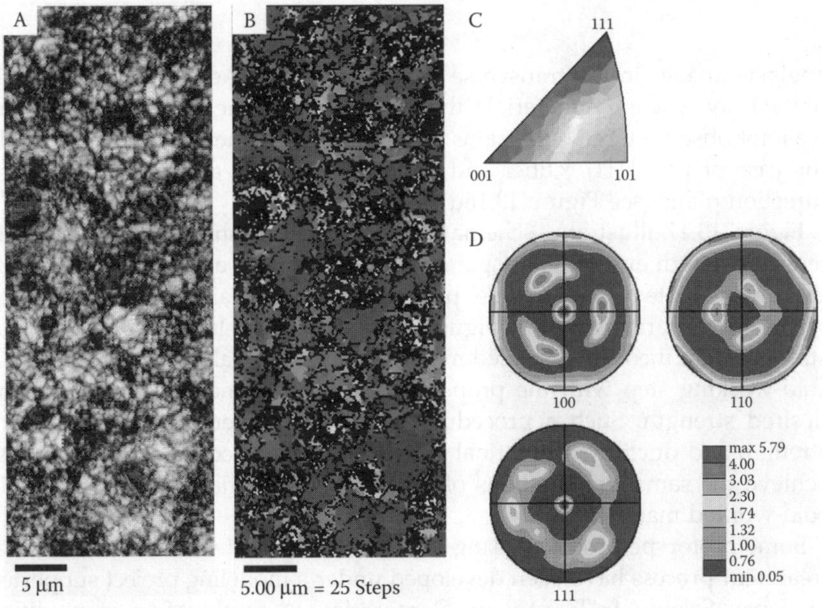

FIGURE 13.16
EBSP analysis images of cold-swaged ($\eta = 2$) and aged at 323 K for 86.4 ks repeated thermomechanical treatment S1 material in the transverse section: (a) image quality map, (b) inverse pole figure (IPF) mapping, (c) color reference of IPF, and (d) texture.

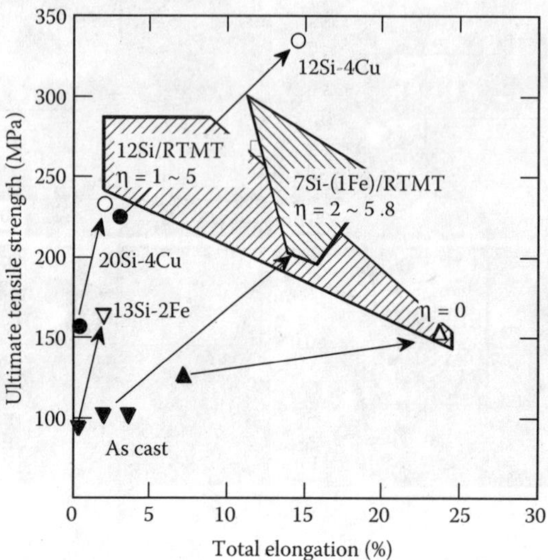

FIGURE 13.17

Illustration of strength–elongation balance map for as-cast and repeated thermomechanical treatment materials.

analysis images in the transverse section of the worked rod. In the matrix of the heavily worked materials, the high density of the dislocation structure was not observed, but sub-grains ca. 200 nm in diameter were detected. In the case of rod, {111} γ fiber and {100} cube texture appear in the rolling direction plane (see Figure 13.16d).

Figure 13.17 illustrates schematically the relationship between ultimate tensile strength and total elongation within wrought alloys.[4–6] It is obvious that an adequate balance of these properties can be obtained within the range of the trend band shown in Figure 13.17. If the finished product must be stronger than the fully annealed material, then the final operation must be a cold-working step with the proper degree of deformation to produce the desired strength. Such a procedure to develop a certain combination of strength and ductility in the final product is more successful than trying to achieve the same combinations of properties by partially softening a fully cold-worked material.

Some prototype products using the modified repeated thermomechanical treatment process have been developed under a modeling project supported by Japan Science & Technology Corporation. A combination of modified repeated thermomechanical treatment using drawing-annealing steps and isothermal warm-forging on Al-Si based alloys has been applied in the manufacturing of a motorbike engine piston. The product benefits by virtue of less machining, and is lighter in weight.

Summary

A new design concept for a recyclable material has been proposed where a combination of multiple phases with fine microstructure is considered. A novel repeated thermomechanical treatment, has been successful in achieving microstructural refinement of hyper-eutectic Al-Si-(Fe, Cu) cast materials leading to improved tensile properties. It raises the possibility of designing materials with balanced properties via cold-working. A modified repeated thermomechanical treatment of Al-Si based alloys has been applied to develop automotive forged parts, such as engine pistons.

References

1. Halada, K., and R. Yamamoto, *MRS Bulletin*, Nov., 871–879, 2001.
2. For an example, Umezawa, O., and K. Nagai, *Trans. of MRS Japan*, Vol. 20, 190–193, 1996.
3. Umezawa, O., and K. Nagai, Japan Patent Nos. 3005672 (1999), 3005673 (1999), and 3111214 (2000), in Japanese.
4. Umezawa, O., and K. Nagai, *Metall. Mater. Trans. A*, Vol. 30A, No. 8., 2221–2228, 1999.
5. Lim, C. Y., O. Umezawa, and K. Nagai, *Metals and Materials*, Vl. 4, No. 5, 1027–1031, 1998.
6. Umezawa, O., H. Yokoyama, and K. Nagai, "Advances in Materials Engineering and Technology," *Int. J. of Materials and Product Technology*, Special Issue, SPM1, Vol. 2, 568–573, 2001.

14

Stainless Steel Sandwich Sheets with Fibrous Metal Cores

A. E. Markaki and Bill Clyne

CONTENTS

Introduction

It has long been recognized that sandwich structures, composed of stiff outer layers held apart by a low density core, offer the potential for very high specific stiffness and other attractive mechanical properties. Most such structures are created by an assembly step of some sort, before or during component manufacture. This limits the flexibility of the production process and is relatively expensive. Nevertheless, there is considerable current interest in sandwich structures of different types, many of them based on metallic faceplates and having metal-containing cores—often made of stochastic cellular metals,[1–3] or some more regular structure such as a truss assembly.[4–6] However, traditional approaches to the fabrication of the latter (e.g., investment casting) are cumbersome and economically unattractive. A particular case of cellular metals is those made of a metallic fiber network of some sort. A novel type of a sandwich steel sheet with a fibrous stainless steel core has recently been developed[7] based on a pair of thin (~200 μm) stainless steel faceplates. This material has been termed a hybrid stainless steel assembly (HSSA). It can, in principle, offer a highly attractive set of property combinations, such as low areal density, high beam stiffness, efficient energy absorption during crushing, and good vibrational damping capacity. Furthermore, the overall thickness of the sheet (~1 mm), and certain features of the core structure, are such that its processing characteristics can be comparable with those of a conventional monolithic metallic sheet. Some work has recently been published on the elastic properties[8] and interfacial delamination behavior[9] of such material.

In addition to mechanical stiffness and strength, electrical properties are relevant, since efficient fabrication using steel sheets requires that they should be weldable, preferably by electrical resistance (spot) welding. This is considered to be highly desirable for use in the automotive industry.[10] There has been extensive work on simulation of heat and current flow during spot welding[11–14] on applied force characteristics,[15] and on welding of thin monolithic metallic sheets.[16] However, there is only limited information available[17] on welding of metallic sandwich sheets with fibrous metallic cores, although there has been some work[18] on welding of vibration-damped steel. This consists of two mild steel sheets, typically 0.3 to 1 mm thick, separated by a thin layer of polymeric adhesive (~20 to 500 μm).

In this chapter, a study is presented of the mechanical and electrical properties of three different variants of hybrid stanless steel assembly material, and of their welding characteristics. Experimental data are correlated with predictions based on simple analytical treatments, and some conclusions are drawn about the advantages and disadvantages of the variants concerned.

Experimental Procedures

Material Production

Three core structures have been investigated: (a) transversely aligned fibers bonded to the faceplates by adhesive (designated "flocked sheet"), (b) in-plane pre-fabricated sintered mesh bonded to the faceplates by adhesive or brazing ("long fiber in-plane mesh"), and (c) 3-D brazed fiber array ("short fiber 3-D array") brazed to the faceplates. In all three cases, the faceplates were made of 200 µm thick 316L austenitic stainless steel. Manufacturing procedures are briefly outlined below:

(a) **Flocked sheet**. This is made by a flocking process,[7] in which short (~1 mm) drawn 316L stainless steel fibers, about 25 µm in diameter, are approximately transversely aligned (see Figure 14.1a). The fibers have an austenitic/martensitic microstructure, giving them high strength, but relatively low ductility. The fibers are adhesively bonded to the faceplates using a two-component epoxy adhesive (Araldite® 420A/B). The fiber volume fraction in the core is about 10%.

(b) **Long fiber in-plane mesh**. This is made by bonding to the faceplates a pre-manufactured solid-state sintered mesh (Bekeart S.A.), containing 19 vol.% of 25 µm diameter fibers of 316L stainless steel, about 16 mm in length. Most of the fibers in the mesh are inclined at a relatively high angle (~80°) to the vertical. The starting fibers are the same as those used in (a), but the heat treatment involved during sintering leads to complete conversion of the martensite to austenite, lowering the strength, and raising the ductility. Bonding of the core to the faceplates was carried out either by adhesive bonding, using a two-component epoxy adhesive (Araldite® 420A/B), or by brazing, using an Ni-14Cr-4 Fe-2.8B-3.3Si-0.6C braze alloy (Brazing & Soldering Automation Ltd.) and a brazing temperature of 1100°C for a period of about 5 minutes.

(c) **Short fiber 3-D array**. This consists of a 3-D network of fibers bonded to each other, and also to the faceplates, using the same braze alloy as for the long fiber in-plane mesh. In this case, the fibers are inclined at various angles to the vertical, with an average value of the order of 60°. They occupy about 10% of the volume of the core, and are made of 446 (ferritic) stainless steel, melt extracted to lengths of 2.5 mm (product of FibreTech Ltd.). These fibers have high strength, combined with limited ductility.

Through-Thickness Stiffness

The through-thickness Young's moduli of the sheets were measured using the high load head of a Micromaterials NanoTest 600 indenter. A flat-ended

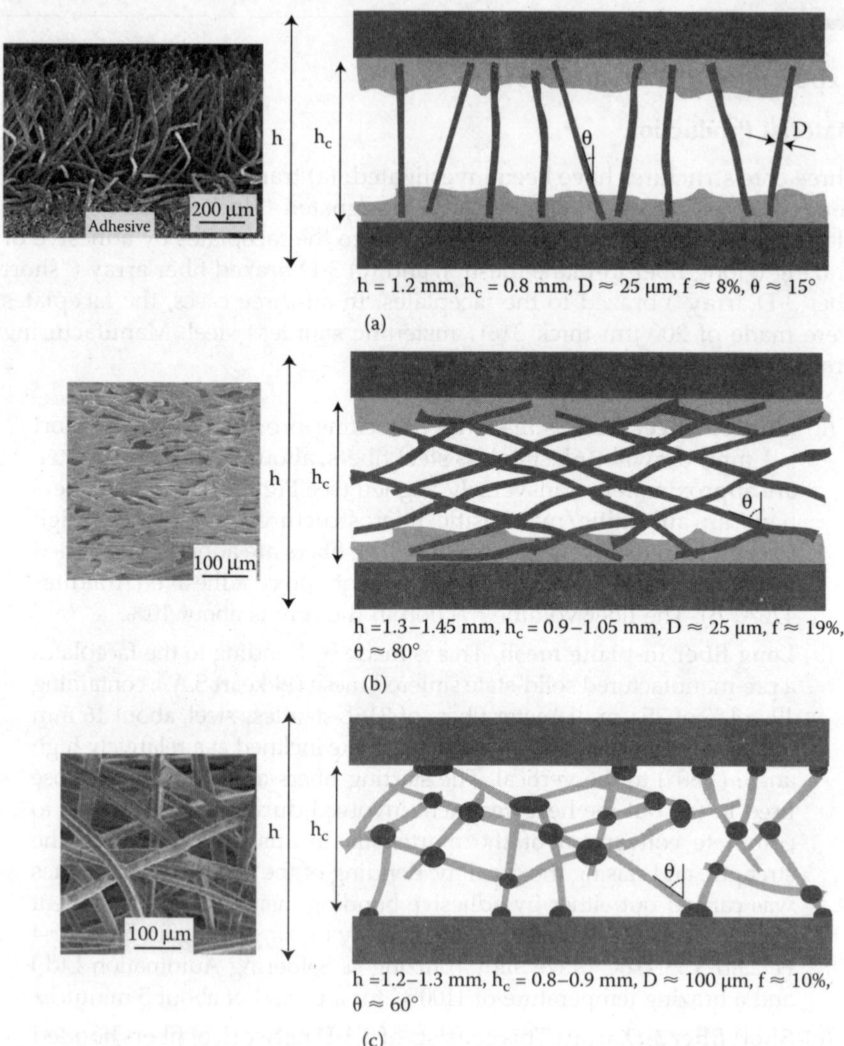

FIGURE 14.1
SEM micrographs showing cross-sectional view of the core and schematic depictions of the structure of (a) flocked sheet, (b) long in-plane sheet, and (c) short fiber 3-D array.

cylindrical indenter head was used, with a diameter of 25 mm. The specimens were cut to square sections with sides of length about 5 mm. Loads of up to about 20 N were applied, corresponding to average stresses on the faceplate of around 0.8 MPa.

Through-Thickness Yield Stress

Small coupons, measuring 10 mm by 15 mm, were loaded in compression on a servo-hydraulic testing apparatus equipped with a 10 kN load cell.

The tests were conducted under controlled displacement rate of 0.15 mm.min^{-1}, and the through-thickness displacements were recorded from an LVDT ($\pm$ 500 μm range) with a resolution of about 5 μm. Attention was focused on identification of the core yield stress, corresponding to the onset of substantial plastic bending and buckling among the fibers in the core.

Interfacial Fracture Energy

The resistance to delamination of the faceplates was measured under mode I loading conditions. The test procedure employed[19] involves bonding of the sheet to two steel plates and loading these under pure bending, so as to generate mode I crack growth. Prior to testing, a narrow pre-crack was introduced into the specimen at the mid-plane.

Single Fiber Tensile Testing

Single fiber testing was carried out using a Schenk desktop testing machine, fitted with a 5 or 250 N load cell. The individual fibers were mounted on paper tabs, with a central cut-out that gave a gauge length of about 25 mm. The tab was gripped in the jaws of the testing machine; prior to testing, cuts were made from each side to the central cut-out. The cross-head displacement was measured using an LVDT. All tests were conducted in displacement control at a rate of 0.1 mm.min^{-1}. Fibers were tested in the as-received condition, and also after a heat treatment similar to those involved during solid-state sintering or brazing.

Electrical Conductivity

Specimens were in the form of small rectangular coupons (10 $\times$ 14 mm). An AC circuit was used for the measurements. A fixed current (1 A) was passed through the specimen via flat-ended probes on the faceplates. This current was modulated at a constant frequency of 1 kHz. The potential drop across the specimen, from which the resistivity of the core was deduced, was then amplified and measured. In making this measurement, the potential drop across other resistances in the sensing circuit must be eliminated. This drop generates an offset in the mean voltage of the AC signal. This DC offset was removed by passing the signal through an AC amplifier, after which it was rectified by a demodulator, passed through an integrator to remove noise, and finally, displayed as a DC voltage (few mV to $\sim$ 1 V for these specimens). Full details of the technique are given elsewhere.[20]

Resistance Welding

Resistance welding trials were conduced on a single phase AC machine. Cu/Cr/Zr electrodes to ISO 5821 type E design were used, having 16 mm diameter and 6 mm radius dome tip. An electrode force of 1.5 kN was used.

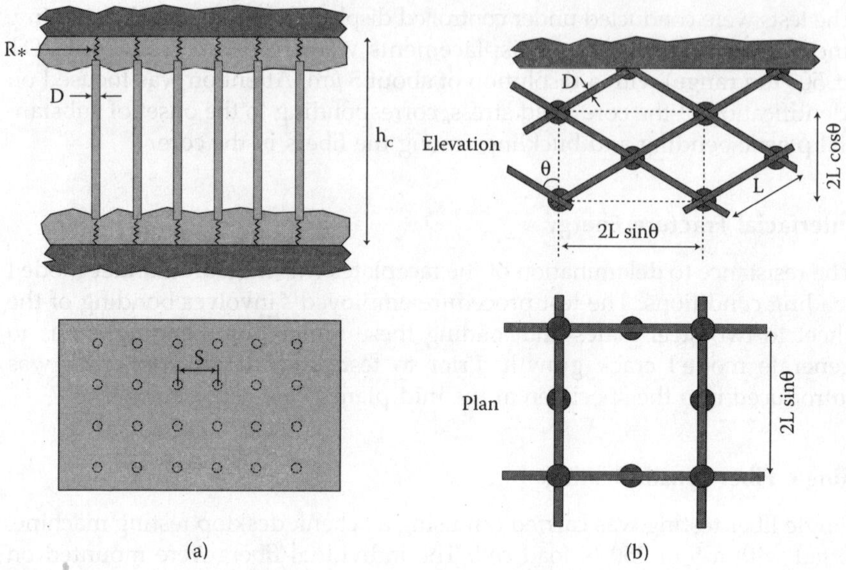

FIGURE 14.2
Schematic showing modeled fiber distributions for (a) the transverse fiber and (b) the sintered mat (right) structures.

Single welds were made on small coupons. In cases where the core did not allow significant currents to flow at the low voltages used in resistance welding (typically about 5 V), a clamp tool was used to provide a current shunt path (in parallel with the electrodes) linking the outer sheet surfaces. This current flow heated the specimen sufficiently to allow deformation of the core, and hence, direct flow through it, so that welding could be achieved. The welding current and voltage were measured using a standard meter and oscilloscope.

Geometrical Representation of the Core Structure

Simple analytical models have been developed, giving the fiber volume fraction f as a function of geometrical variables. For the flocked sheet core, a square array was assumed, with side of length S (Figure 14.2a). The fiber volume fraction is thus given by

$$f = \frac{\pi D^2}{4 S^2} \tag{14.1}$$

where D is the fiber diameter. The number of fibers per unit area, N, is related to the fiber volume fraction f and the fiber diameter, D

$$N = \frac{4f}{\pi D^2} \tag{14.2}$$

For the long fiber in-plane mesh, a tetragonal unit cell is identified, of side $2L.\sin\theta$ and height $2L.\cos\theta$, where L is the length of a segment of fiber having one end at a cell corner and the other at the mid-point of a vertical face (Figure 14.2b). Each cell thus contains 16 fiber segments, all of which are shared between 2 cells. The fiber volume fraction and the number of fibers per unit area are thus given by

$$f = \frac{8\left[L\left(\dfrac{\pi D^2}{4}\right)\right]}{(2L\ \sin\theta)^2(2L\ \cos\theta)} = \frac{\pi D^2}{4L^2\sin^2\theta\ \cos\theta} \qquad (14.3)$$

$$N = \frac{4}{(2L\sin\theta)^2} = \frac{4f\ \cos\theta}{\pi D^2} \qquad (14.4)$$

For the short fiber 3-D array, the relationship between N and f is simply obtained by noting that, for a set of prisms with a 3-D random orientation distribution of the prism axes, the area intersected by any plane is twice the area intersected by a plane lying normal to the alignment direction of a set of parallel prisms occupying the same volume fraction.[21] Hence, N here has a value of half that for the case of an aligned set of cylinders $(=f/(\pi D^2/4))$, i.e.,

$$N = \frac{2f}{\pi D^2} \qquad (14.5)$$

Through-Thickness Loading Response

Core Stiffness

Two different approaches, based on a cantilever bending model, have been used to predict the through-thickness stiffness of the sandwich sheet cores. The first is applicable for the flocked sheet and long fiber in-plane mesh, in which the fibers are inclined at a specified angle, whereas the second approach assumes a three-dimensionally random orientation distribution of the fiber axes (short fiber 3-D array).

Flocked Sheet and Long Fiber In-Plane Mesh

The Young's modulus of the core in the through-thickness direction can be predicted by considering the behavior of a single fiber of length L, initially straight and inclined at an angle θ to the direction of the applied load. The situation under load is depicted schematically in Figure 14.3. All fibers

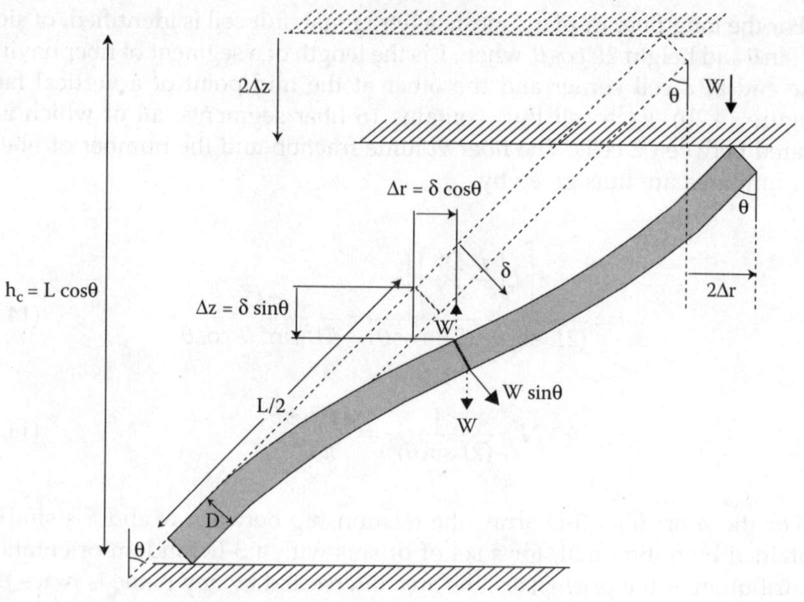

FIGURE 14.3
Elevation view of the elastic bending of an inclined fiber under the influence of a vertical load, *W*.

inclined at such an angle will behave similarly under the action of an imposed load *W* normal to the plane of the sheet, provided any interaction between individual fibers is neglected, and assuming that their behavior remains linearly elastic. From elementary beam bending theory, the normal deflection, δ (=$\Delta z/\sin\theta$), of the free end of a cantilever beam of length $L/2$, subjected to a load $W \sin\theta$ normal to the beam axis, is given by

$$\frac{\Delta z}{\sin\theta} = \frac{1}{3}\frac{W \sin\theta \,(L/2)^3}{E_f I} \tag{14.6}$$

where E_f is the fiber modulus and I is the moment of inertia of the fiber section (=$\pi D^4/64$, where D is the fiber diameter, for cylindrical fibers).

The through-thickness strain of the core, ε_c, is given by

$$\varepsilon_c = \frac{2\Delta z}{h_c} \tag{14.7}$$

Substituting for Δz from Equation 14.6, and writing the core thickness as $L \cos\theta$ (see Figure 14.3), leads to

$$\varepsilon_c = \frac{2W \sin^2\theta \,(L/2)^3}{3E_f\,(\pi D^4/64)\, L \cos\theta} = \frac{16W \sin^2\theta \, L^2}{3E_f\,\pi D^4 \cos\theta} \tag{14.8}$$

Now, the applied pressure, P, can be expressed in terms of the value of W and the number of fibers per unit area, N

$$P = NW \tag{14.9}$$

Substituting for N from Equation 14.4, the Young's modulus of the core in the through-thickness direction can be expressed as

$$E_c = \frac{P}{\varepsilon_c} = \frac{3NW \, E_f \, \pi D^4 \cos\theta}{16 \, W \sin^2\theta \, L^2} = \frac{3 \, (4f \cos\theta) \, E_f \, \pi D^4 \cos\theta}{16 \, (\pi D^2) \, \sin^2\theta \, L^2} = \frac{3E_f f}{4 \, s^2 \tan^2\theta} \tag{14.10}$$

where $s \ (=L/D)$ is the fiber aspect ratio.

Predictions* obtained using Equation 14.10 show that higher stiffness is predicted for fibers inclined at low angles, as expected, since a fiber provides less resistance to vertical displacement when it is inclined at a high angle. It may also be noted that, for a given volume fraction of fiber, there will be more fibers per unit area when θ is close to 0 (see Equation 14.4). Also, the fibers will be of lower aspect ratio, for a given core thickness. The net effect is cumulative, so the sensitivity of the stiffness to θ is quite strong ($\tan^2\theta$). Furthermore, it can be seen that the stiffness goes up sharply as the fiber aspect ratio is decreased. An obvious way of increasing the stiffness, for a given core thickness, is to use fibers with larger diameter. An increase in fiber content will also generate increased stiffness, but this is less efficient and, of course, it also raises the density of the core.

3-D Random Fiber Array

The cantilever bending model (Figure 14.3) can be also used to predict the elastic stiffness of an isotropic random fiber array. In this case, the fiber being considered does not span the two faceplates, but is just a segment between two fiber-fiber joints. The deflection is induced by an applied stress, σ (compressive in Figure 14.3), which generates a force W on each individual fiber segment. These are related by

$$\sigma = NW \tag{14.11}$$

where N is the number of fiber segments per unit sectional area.

---

* This treatment clearly breaks down in the limit of $\theta=0$, when the predicted stiffness tends to infinity. The value must be upper bounded at $f \, E_f$, corresponding to the fibers remaining vertical and being axially compressed. In practice, even this value would not be approached, at least for fibers with relatively high aspect ratios. Realistically, the model may be taken as applicable for angles down to around 5–10, which is probably all that is required.

As before, the normal deflection, δ, of the end of a cantilever beam of length $L/2$, subjected to a load $W\sin\theta$ normal to the beam axis, is given by

$$\delta = \frac{W\sin\theta\,(L/2)^3}{3E_f\left(\dfrac{\pi D^4}{64}\right)} = \frac{8W\sin\theta\,L^3}{3E_f\pi D^4} \tag{14.12}$$

so that, substituting for W from Equations 14.11 and 14.12 and N from Equation 14.5, the axial deflection is given by

$$\Delta z = \delta\sin\theta = \frac{4\sigma L^3\sin^2\theta}{3E_f\,f\,D^2} \tag{14.13}$$

The overall deformation expected with a material composed of a three-dimensionally random array of fibers can be analyzed, at least approximately, by summing the contributions* from individual fiber segment deformations. The segments are assumed to exhibit a spherically symmetric orientation distribution, which has a $\sin\theta$ angular probability distribution about any given axis. The expected overall relative net extension in the direction of an applied stress can, therefore, be written down by considering the displacements of a set of fiber mid-points, using the expression for the deflection normal to the fiber axis, as a function of the distance along the fiber, given by Equation 14.13. The macroscopic deflection in the loading direction, and hence the strain, can thus be expressed as

$$\varepsilon_c = \frac{\Delta Z}{Z} = \frac{\displaystyle\int_0^{\pi/2}\Delta z\sin\theta\,\mathrm{d}\theta}{\displaystyle\int_0^{\pi/2} z\sin\theta\,\mathrm{d}\theta} = \frac{\displaystyle\int_0^{\pi/2}\frac{4\sigma L^3\sin^3\theta}{3E_f\,f\,D^2}\,\mathrm{d}\theta}{\displaystyle\int_0^{\pi/2}\left(\frac{L}{2}\cos\theta\right)\sin\theta\,\mathrm{d}\theta}$$

$$\therefore\,\varepsilon_c = \left(\frac{8\sigma}{3E_f\,f}\right)\left(\frac{L}{D}\right)^2\frac{\displaystyle\int_0^{\pi/2}\sin^3\theta\,\mathrm{d}\theta}{\displaystyle\int_0^{\pi/2}\cos\theta\sin\theta\,\mathrm{d}\theta} = \left(\frac{32\sigma}{9E_f\,f}\right)\left(\frac{L}{D}\right)^2 \tag{14.14}$$

---

* In reality, the deflections exhibited by individual fiber segments will be influenced by the configuration of neighboring segments, so this analysis is clearly not rigorous when the inclination angles vary within the material. For example, the axial deflection of a segment inclined at a substantial angle to the stress axis would be reduced if a closely neighboring segment were aligned parallel to the axis. However, such interactions are unlikely to generate large errors in the proposed model, at least for an effectively isotropic, homogeneous material.

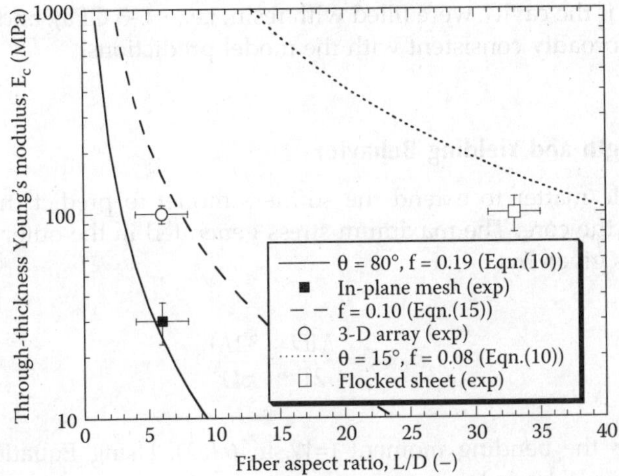

FIGURE 14.4

Through-thickness Young's modulus of the core as a function of fiber aspect ratio. The three points correspond to experimental measurements, while the curves are predictions obtained using Equations 14.10 and 14.15. A value of 200 GPa was used for the stiffness of the stainless steel fibers.

The Young's modulus of the fiber array, E_c ($=\sigma/\varepsilon$), is therefore given by

$$\therefore E_c = \frac{9E_f f}{32\left(\dfrac{L}{D}\right)^2} = \frac{9E_f f}{32}\frac{1}{s^2} \tag{14.15}$$

A comparison between experimental results and predictions obtained using Equation 14.10 (flocked sheet, long fiber in-plane mesh) and Equation 14.15 (short fiber 3-D array) is shown in Figure 14.4. It can be seen that experimental data are broadly consistent with predictions from the model predictions. For the flocked sheet, the measured value of about 100 MPa is in fairly good agreement with predictions obtained using estimated values for s, f and θ of 33 ($= h_c /D.\sin\theta$), 8% and 15°, respectively. For the in-plane mesh, stiffnesses of the order of 30 MPa were obtained experimentally. Using estimated values for s, f, and θ of 6, 19%, and 80°, respectively gives a predicted stiffness of about 25 MPa. For cores in which the fibers are bonded together, the appropriate aspect ratio to use is that between fiber joints: this value can be estimated from study of SEM micrographs, such as that shown in Figure 14.1b. For the short fiber 3-D array, the measured value of about 100 MPa is in fairly good agreement with predictions obtained using estimated values for s and f of 6% and 10%, respectively. In any event, it is clear that all of these stiffnesses are relatively low (appreciably lower, f or example, than would be the case if the cavity were filled with resin, i.e., ~1–3 GPa), and that these results are broadly consistent with the model predictions.

Core Strength and Yielding Behavior

It's a simple matter to extend the stiffness model to predict the onset of yielding in the core. The maximum stress generated in the outer surface of the fibers is given by

$$\sigma_{f,max} = \frac{MD}{2I} = \frac{32M}{\pi D^3} \tag{14.16}$$

where M is the bending moment ($=W.\sin\theta.L/2$). Using Equation 14.8 to substitute for W leads to

$$\sigma_{f,max} = \frac{32\,(3E_f\,\pi D^4\cos\theta\,\varepsilon_c)\sin\theta\,L/2}{\pi D^3\,(16\sin^2\theta\,L^2)} = \frac{3E_f\,\varepsilon_c}{s\tan\theta} \tag{14.17}$$

Equation 14.17 can be used to obtain an expression for the yield stress of the fiber, $\sigma_{f,Y}$, in terms of the through-thickness yield strain of the core $\varepsilon_{c,Y}$

$$\sigma_{f,Y} = \frac{3E_f\,\varepsilon_{c,Y}}{s\tan\theta} \tag{14.18}$$

The through-thickness yielding stress of the core $\sigma_{c,Y}$, in turn, can be expressed as

$$\sigma_{c,Y} = E_c\,\varepsilon_{c,Y} \tag{14.19}$$

Substituting for E_c and $\varepsilon_{c,Y}$ from Equations 14.10 and 14.18, respectively, $\sigma_{c,Y}$ can be expressed in terms of $\sigma_{f,Y}$

$$\sigma_{c,Y} = \frac{3E_f\,f}{4\,s^2\tan^2\theta}\,\frac{\sigma_{f,Y}\,s\tan\theta}{3E_f} = \frac{f\,\sigma_{f,Y}}{4\,s\tan\theta} \tag{14.20}$$

From Equation 14.20, it can be seen that the core yielding stress rises as the fiber aspect ratio is decreased, although the dependence is not as strong as that exhibited by the stiffness. A plot is shown in Figure 14.5 of predictions from this equation for the short fiber 3-D array ($s = 6$, $f = 0.10$, $\sigma_{f,Y} = 1000$ MPa), the long fiber in-plane mesh ($s = 6$, $f = 0.19$, $\sigma_{f,Y} = 200$ MPa), and the flocked sheet ($s = 33$, $f = 0.08$, $\sigma_{f,Y} = 1000$ MPa). The yield strengths of the heat-treated 446 (short fiber 3-D array), heat-treated 316L (in-plane mesh),

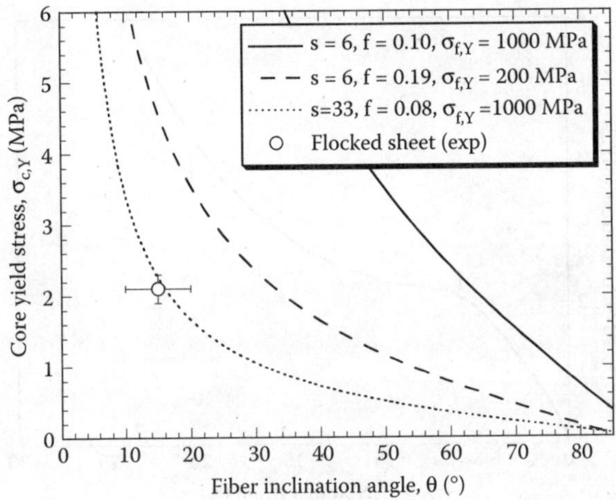

FIGURE 14.5

Predicted dependence (Equation 14.20) of the through-thickness yield stress of the core on the fiber inclination angle. The yield strength of the fibers was obtained from the stress–strain curves of Figure 14.6. Also plotted is the experimental result for the yield stress of the flocked core obtained from compression testing (Figure 14.7).

and the as-received (drawn) 316L (flocked sheet) fibers were obtained from single fiber tensile testing (see Figure 14.6). Also plotted in Figure 14.5, is the experimentally measured value of $\sigma_{c,Y}$ obtained from compression of the flocked sheet, which can be inferred from Figure 14.7 to be about 2.1 MPa. It can be seen that this value is in good agreement with the model predictions.

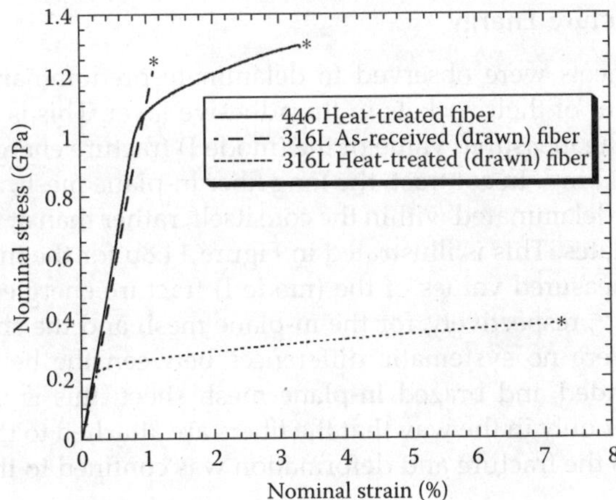

FIGURE 14.6

Typical single fiber tensile testing data for (a) as-received (drawn) 316L (flocked sheet), (b) heat-treated 316L (long fiber in-plane mesh), and (c) heat-treated 446 (short fiber 3-D array) fibers.

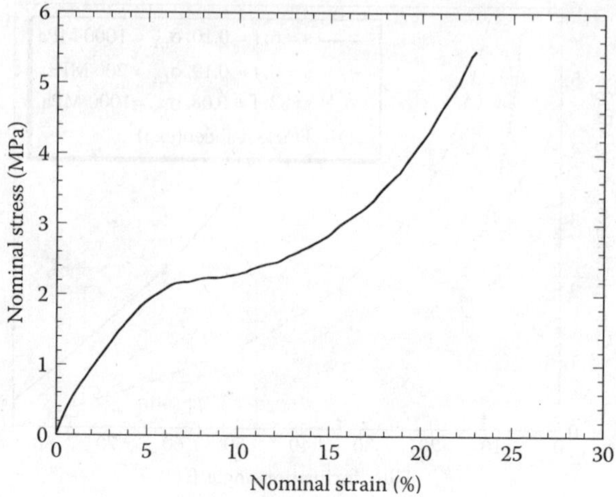

FIGURE 14.7
Typical stress-strain plot for compressive testing of the flocked sheet.

For the other two materials, on the other hand, it is difficult to quantify this value experimentally, at least during compression testing, because extensive yielding only occurs after substantial densification has taken place, so the compressive stress-strain curve does not display a distinct plateau regime. In practical terms, it was evident on general handling and testing of the sheets that the short fiber 3-D random core material is substantially stronger than the other two.

Interfacial Fracture Energy

The flocked sheets were observed to delaminate predominantly by fibers being pulled out of their sockets in the adhesive layer. This is illustrated in Figure 14.8a. The measured value of the (mode I) fracture energy was found to be about 340 J m^{-2}. In contrast, the long fiber in-plane mesh, and the short fiber 3-D array delaminated within the core itself, rather than at the interfaces with the faceplates. This is illustrated in Figure 14.8b for the in-plane mesh. The average measured values of the (mode I) fracture energies were about 30 and 675 J m^{-2}, respectively, for the in-plane mesh and the short fiber 3-D array. There were no systematic differences between the behavior of the adhesively bonded and brazed in-plane mesh sheet; this is unsurprising, since they differ only in the way that the fibers are attached to the faceplates, and in all cases the fracture and deformation was confined to the mid-plane region.

However, there is a substantial difference in the fracture energies between the in-plane mesh and the short fiber 3-D array. This may be attributed to the different bonding techniques employed to generate joints between adjacent fibers.

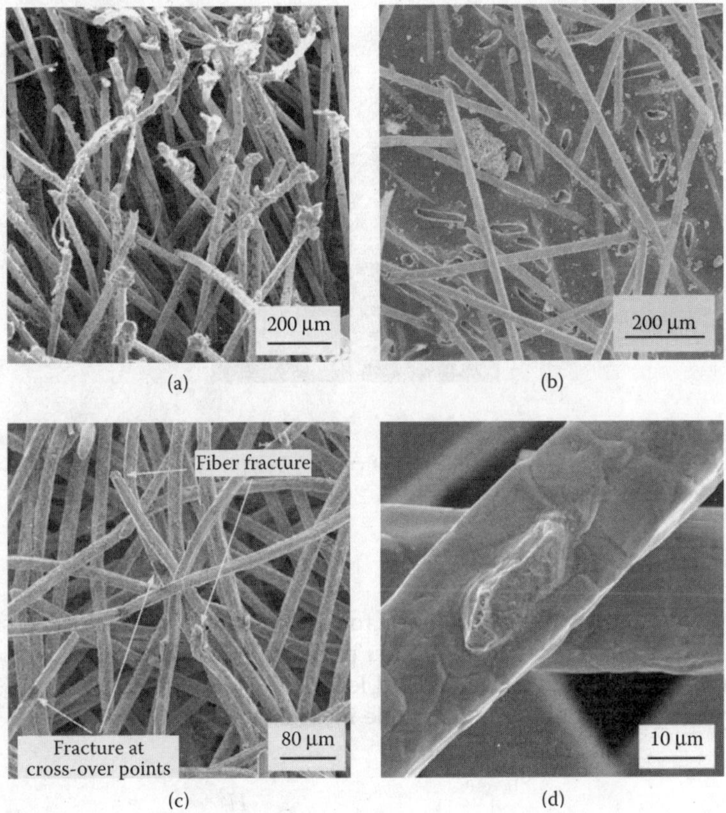

FIGURE 14.8
Scanning electron micrographs of fracture surfaces. The top pair are from a flocked sheet specimen showing (a) glue adhering to the ends of pulled-out fibers and (b) sockets in the adhesive layer, from which fibers had been pulled out. The bottom pair are from a long fiber in-plane mesh showing (c) a general view and (d) a fractured neck formed at a fiber-fiber joint.

Solid-state sintering is a very slow process and, even after prolonged holding at high temperature, the joints are likely to be limited in area, and consequently weak. Furthermore, the prolonged heating can lead to grain coarsening and extensive carbide precipitation, reducing the fiber strength. On the other hand, liquid phase sintering techniques, such as brazing, are much faster than solid state sintering, since the rate-determining process is viscous flow, rather than diffusion. Well-consolidated joints can, therefore, be formed quickly and readily, with less danger of deleterious changes in the fiber microstructure. Of course, the mechanical properties of the braze metal may be relevant, but with suitable joint geometry the stresses in the braze metal will be much lower than those in the fibers, making it unlikely that the joints will fail.

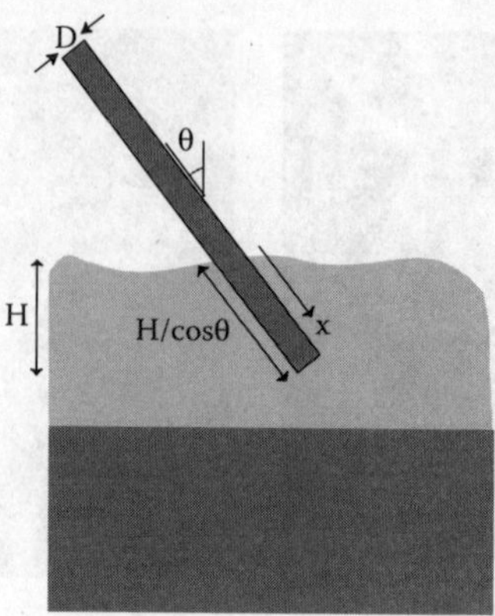

FIGURE 14.9
Schematic of the model used for prediction of energy absorption during pull-out of inclined fibers for the flocked sheet core.

Fiber Pull-Out in Flocked Sheets

A simple model has been developed for delamination in the flocked sheet, based on a shear-lag treatment of fiber pull-out.[22] Consider an inclined fiber with a diameter D and an embedded length x, as illustrated in Figure 14.9. The work done in pulling out a single inclined fiber can be written as

$$\Delta G = \int_0^{H/\cos\theta} \pi D x\, \tau_{i^*}\, dx = \pi D \frac{H^2}{2\cos^2\theta}\, \tau_{i^*} \tag{14.21}$$

where τ_{i^*} is the fiber/adhesive interfacial shear strength, taken as constant along the fiber length. If the number of fibers per unit area is written as N, there will be $(N\cos\theta\, dx\, /H)$ per unit area with an embedded length between x and $(x + dx)$. Thus the total work done in pulling out the inclined fibers, G_{fp} is given by

$$G_{fp} = \int_0^{H/\cos\theta} \frac{N\cos\theta\, dx}{H} \pi D \frac{H^2}{2\cos^2\theta}\, \tau_{i^*} \tag{14.22}$$

Substituting the expression for N given in Equation 14.4 into Equation 14.22 and integrating (assuming equal embedded lengths for all fibers) leads to

$$G_{fp} = \frac{2f\, \tau_{i^*}\, H^2}{D\cos\theta} \tag{14.23}$$

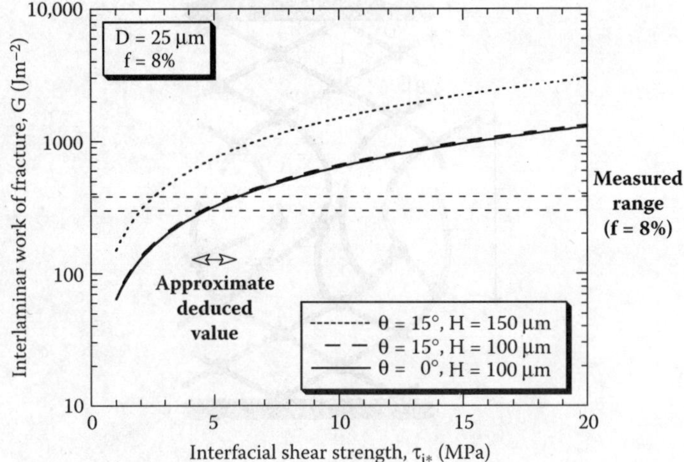

FIGURE 14.10
Predicted dependence (Equation 14.23) of the interlaminar fracture energy of the flocked sheet on the fiber-adhesive interfacial shear strength.

Figure 14.10 shows the predicted dependence of this pull-out energy on interfacial shear strength. From SEM images of cross-sections through the sheet, it was estimated that the fibers are typically anchored into the adhesive to a depth of about 100 μm ($=H$). It can be seen that the experimentally obtained value of G ($\sim$340 J.m^{-2}) corresponds to that predicted by the model (using $f = 8\%$, $H = 100$ μm, $\theta = 15$ and $D = 25$ μm) if the fiber/adhesive interfacial shear strength has a value of about 5 MPa. This is a relatively low value for an interfacial shear strength,[23,24] although it is certainly of the order of magnitude expected.

Fiber Fracture in the In-Plane Mesh and 3-D Array

The long fiber in-plane mesh and the short fiber 3-D array were observed to delaminate by fiber fracture. A simple model has been developed to estimate the fracture energy for this mechanism. It is assumed that all the fibers deform and fracture within a deformation zone of length z. This is illustrated in Figure 14.11. The work of fracture may, in this case, be written as

$$G_{\text{fr}} = N U_{\text{s}} z \qquad (14.24)$$

where U_{s} is the work of fracture for a single fiber. Substituting the expression for N given in Equation 14.4 into Equation 14.24 leads to an expression for the work of fracture of the long fiber in-plane mesh

$$G_{\text{fr}} = \left[\frac{4 f \cos\theta}{\pi D^2} \right] U_{\text{s}} z \qquad (14.25)$$

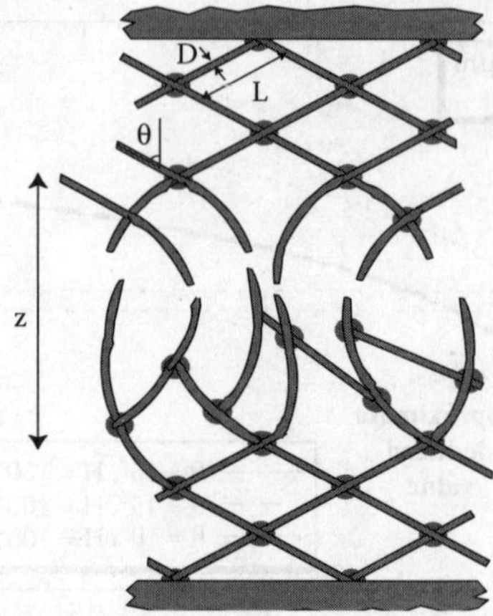

FIGURE 14.11
Schematic of the model used for prediction of energy absorption during fiber fracture for the in-plane mesh and the 3-D array.

Predictions obtained using this equation are shown in Figure 14.12. The work of fracture for a heat-treated 316L fiber was measured from the area under the load-strain curve (Figure 14.6), and is approximately 0.0011 $J.m^{-1}$. The experimental value of G (~30 $J.m^{-2}$) is consistent with model predictions if the fibers are inclined at 85° to the stress axis, and the deformation zone is about 100 μm long. Evidently, the length of the deformation zone is an important parameter, and these results suggest that deformation is restricted

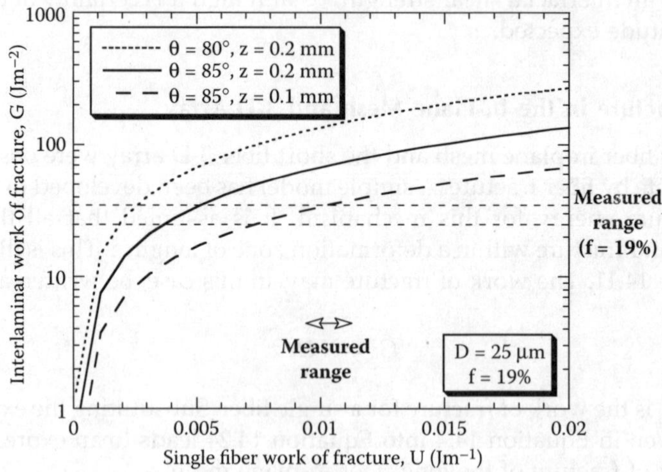

FIGURE 14.12
Predicted dependence (Equation 14.25) of the interlaminar fracture energy of the long in-plane mesh on the single fiber work of fracture.

to a fairly narrow band. This is not something that can readily be verified by inspection of failed specimens, since it is rather difficult to establish precisely where substantial plastic deformation of the fibers has occurred. Nevertheless, the general impression on inspecting the damage and deformation zone is that it was wider than 100 μm. Moreover, the value of 85° used for the fiber inclination angle is probably too high.

It can thus be seen that the experimental value is actually considerably lower than that predicted using the model. The probable explanation for this is evident in Figure 14.8c and 14.8d, where it can be seen that failure commonly occurred at the sintered necks, rather than by fiber fracture. As illustrated in Figure 14.8d, the width of the necks can be quite small, relative to the fiber diameter. Consequently, when the faceplates are torn apart, necks that are not sufficiently strong are apparently quite prone to fail. This probably accounts for the measured fracture energy values being lower than predicted.

For the short fiber 3-D array, substituting the expression for N given in Equation 14.5 into Equation 14.24 leads to the following expression for the work of fracture

$$G_{fr} = \left[\frac{2f}{\pi D^2} \right] U_s\, z \qquad (14.26)$$

Predictions obtained using this equation are shown in Figure 14.13. The work of fracture for a heat-treated 446 fiber was measured from the area under the load-displacement curve (Figure 14.6), and is approximately 0.2 J.m^{-1}. The experimental value G (~ 675 J.m^{-2}) broadly agrees with model predictions if the deformation zone is taken to have a length of the order of

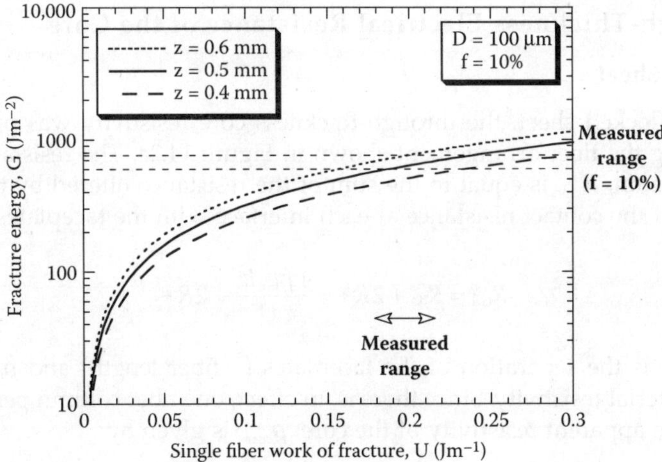

FIGURE 14.13

Predicted dependence (Equation 14.26) of the interlaminar fracture energy of the 3-D fiber array on the single fiber work of fracture.

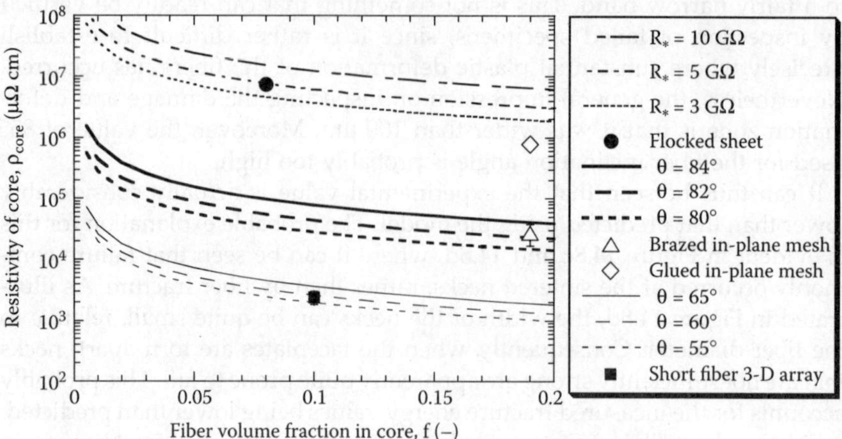

FIGURE 14.14
Core resistivity as a function of fiber content. Points are experimental data, and curves are predictions obtained using Equation 14.29, with three values of $R*$ and $h_c = 0.8$ mm, for the flocked sheet, and Equation 14.32, with three values of θ, for the long in-plane mesh and the short fiber 3-D array. Resistivity values of 85 and 65 $\mu\Omega$.cm, respectively, were used for the austenitic (flocked sheet and long in-plane mesh) and ferritic (short fiber 3-D array) fibers.

0.5 mm (Figure 14.14). It may be noted that this is the approximate spacing between the fiber-fiber joints in this material (see Figure 14.1c).

Through-Thickness Electrical Resistance of the Core

Flocked Sheet

For the flocked sheet, the through-thickness core resistivity was predicted assuming the fiber arrangement shown in Figure 14.2a. The resistance of a fiber column, R_{col}, is equal to the sum of the resistance offered by the fiber itself and the contact resistance at each interface with the faceplates, $R*$

$$R_{col} = R_{fib} + 2R* = \frac{4\,\rho_{fib}\,h_c}{\pi\,D^2} + 2R* \tag{14.27}$$

where h_c is the separation of the faceplates (~ fiber length) and ρ_{fib} is the fiber material resistivity. Since there is, in effect, one fiber column per square array, the apparent resistivity of the core, ρ_{core}, is given by

$$\rho_{core} = \frac{\left(\dfrac{4\,\rho_{fib}\,h_c}{\pi\,D^2} + 2R*\right)S^2}{h_c} \tag{14.28}$$

After substitution for S in terms of f (Equation 14.1), this leads to an expression for the core resistivity in terms of known dimensions, the resistivity of the fiber material, and the contact resistance.

$$\rho_{core} = \frac{1}{f}\left(\frac{R * \pi D^2}{2\,h_c} + \rho_{fib}\right) \tag{14.29}$$

It can be seen that the thickness of the core comes into this expression.

Brazed In-Plane Mesh and 3-D Array

For the brazed in-plane mesh and the short fiber 3-D array, it is assumed that there is no contact resistance at the interface between fibers and faceplates, so the treatment just concerns a small representative volume of core material (Figure 14.2b). The resistance to current flow presented by the unit cell, R_{cell}, can be expressed in terms of the resistance of a segment of fiber R_{seg}. Since, in effect, current passes through two sets of 4 parallel segments as it progresses down the height of a unit cell, it follows that

$$\frac{1}{R_{cell}} = \frac{4}{2R_{seg}} = \frac{2}{\dfrac{L\,\rho_{fib}}{\left(\dfrac{\pi D^2}{4}\right)}} = \frac{\pi D^2}{2\,L\,\rho_{fib}} \tag{14.30}$$

The resistance offered by the unit cell can also be expressed in terms of the resistivity of the core

$$R_{cell} = \frac{\rho_{core}\,2L\cos\theta}{\left(2L\sin\theta\right)^2} = \frac{\rho_{core}\cos\theta}{2L\sin^2\theta} \tag{14.31}$$

Combining Equations 14.3, 14.30, and 14.31 leads to

$$\rho_{core} = \frac{\rho_{fib}}{f\cos^2\theta} \tag{14.32}$$

A comparison between experimental results and predictions (Equations 14.29 and 14.32) is shown in Figure 14.14. It can be seen that, in order to obtain consistency with experimental data, it is necessary to assume a relatively high resistance (few $G\Omega$) between fiber end and faceplate in the flocked fiber core. This can be attributed to poor electrical contact with the faceplates. The experimental resistivities of the brazed in-plane mesh and the short fiber 3-D array cores are orders of magnitude lower than the effective resistivity of the flocked core. The values are consistent with the simple geometrical model, assuming that the fibers are inclined at about 82 and 60, respectively, to the direction of current flow. These figures are at least approximately correct. The resistance of the adhesively bonded in-plane mesh is considerably higher than the predicted levels for a brazed sintered mat, which is attributed to poor electrical contact with the faceplates. However, it can be seen that this is still better than in the flocked material. This is probably because pressure was applied while the adhesive was setting, bringing the fibers into better electrical contact with the faceplates than is possible with the flocked sheet procedure.

Welding Characteristics

The flocked sheet could not be welded directly, since no significant current flowed through the material with the electrode force and voltage used. This is consistent with the high measured electrical resistivity of the core. By using a shunt, however, it was found to be possible to create a weld. Initially, sufficient current flowed through the faceplates and across the shunt to cause heating of the core between the electrodes, leading to softening, consolidation, and hence, sufficient reduction in core resistance to allow a substantial direct current to flow and melting to occur. However, during these initial trials, breakthrough of the core was inconsistent and the faceplates were susceptible to local burn-through by the shunt current. In the example shown in Figure 14.15, breakthrough and current increase occurred only in the last of the 10 cycles of weld time. Even then, the poor shape of the final half cycles of current indicates intermittent current flow. In some cases, depending on the position of the shunt, no breakthrough occurred at all.

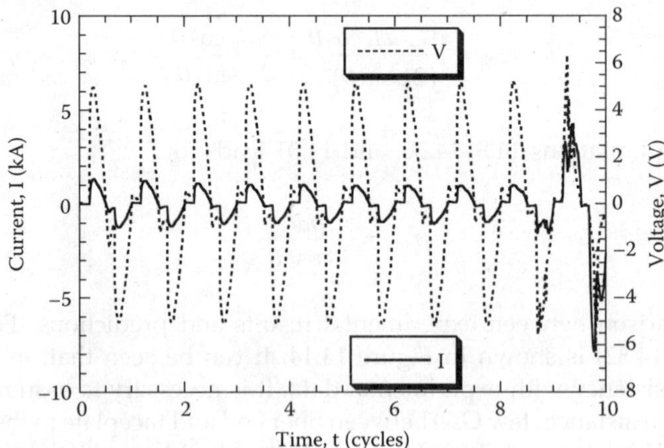

FIGURE 14.15
Voltage-time and current-time plots obtained during welding together of two flocked sheets, with a weld time of 10 cycles (0.2 s).

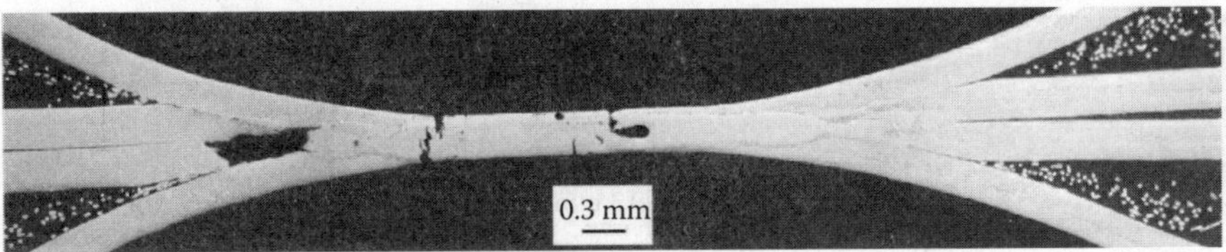

FIGURE 14.16
Optical micrograph of a polished transverse section from a pair of flocked sheets after resistance welding.

Even when a weld was created between flocked sheets, it was invariably of poor quality. This can be seen in the micrograph shown in Figure 14.16, where it is clear that the pressure has resulted in much of the faceplate material being melted and squeezed out laterally. There has also been vaporization of the adhesive, leading to blow-holes, and cracking of the faceplates. Melt expulsion of this type is often problematic, particularly with thin metal sheets in composite materials, such as vibration damping steels.[18] Such a weld would be mechanically very weak.

The brazed long fiber in-plane sheet, on the other hand, was readily weldable. Figure 14.17 shows typical voltage and current plots. The current rises quickly to the set value and substantial heat is generated from the start in the sheets between the electrodes. Sections through a corresponding weld are shown in Figure 14.18. It can be seen that the weld is of good quality, with some lateral flow of melted fibers, but the inner faceplates retaining

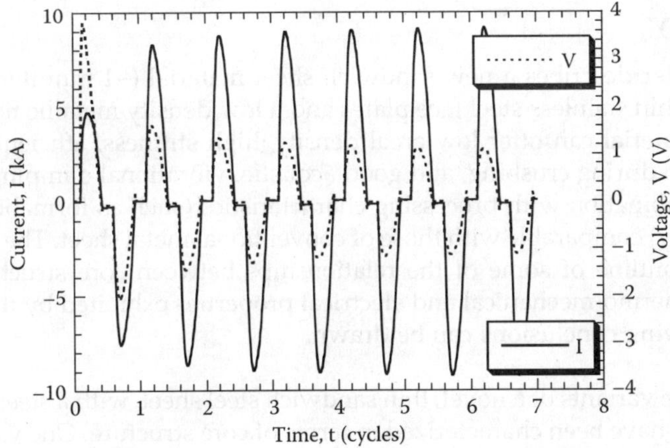

FIGURE 14.17
Voltage-time and current-time plots obtained during welding together of two brazed in-plane mesh sheets, with a weld time of 7 cycles (0.14 s).

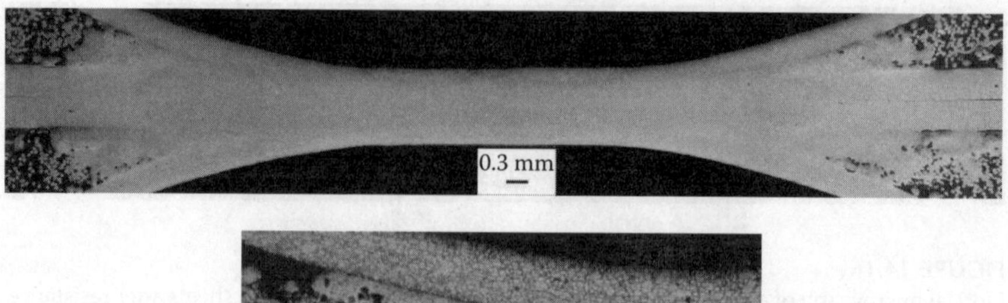

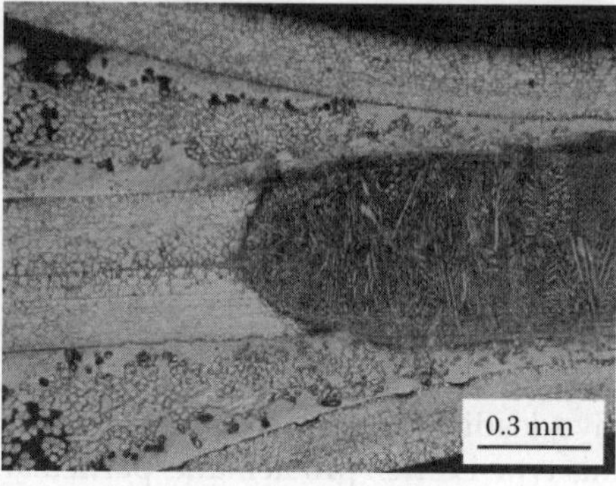

FIGURE 14.18

Optical micrographs of transverse sections from a pair of brazed in-plane mesh sheets after resistance welding, showing an as-polished complete section (above) and a higher magnification view of an etched sample (below), in which the fused zone is clearly visible.

their integrity, and the outer faceplates remaining unmelted. The brazed short fiber 3-D array core material is also readily weldable, as expected.

Summary

This chapter describes a new sandwich sheet material (~1 mm thick) composed of thin stainless steel face plates and a low density metallic fiber core. Such a material can offer low areal density, high stiffness, efficient energy absorption during crushing, and good acoustic/vibrational damping capacity, in combination with processing characteristics (such as formability and weldability) comparable with those of conventional metal sheet. The chapter, gives an outline of some of the relationships between core structure and relevant thermo-mechanical and electrical properties exhibited by the sheet. The following conclusions can be drawn.

(a) Three variants of a novel, thin sandwich steel sheet, with a steel fiber core, have been characterized in terms of core structure. One variant (flocked sheet) contains strong (austenitic/martensitic) fibers oriented approximately normal to the plane of the sheet and bonded to the faceplates by adhesive. The second variant (long fiber in-plane mesh) contains solid-state sintered mats of softer (recrystallized, fully austenitic) fibers oriented at low angles to the plane of the sheet, brazed or adhesively bonded to the faceplates. The third variant (short fiber 3-D array) contains an approximately 3-D random network of strong, melt-spun (ferritic) fibers, brazed to each other and to the faceplates. Fiber contents in the core are around 10% for the flocked sheet and the 3-D array, while the figure is about twice this for the in-plane mesh.

(b) The measured through-thickness Young's moduli are relatively low (~10–100 MPa) and are broadly consistent with predictions from an analytical model based on the bending of individual fibers. Core yield stress values have also been explored, and are in the range 1–10 MPa. In this context, it is worth noting that, in sandwich sheets, the bending stiffness and strength are dominated by the faceplates, so the core does not necessarily need to be very stiff or strong. However, very compliant, or very soft, cores might be problematic, in that they may allow excessive shear between the faceplates or failure to maintain faceplate separation under load, leading to low beam stiffness or premature plastic deformation of the sheet. The brazed 3-D random fiber array core performs appreciably better than the other two in this regard.

(c) The fracture energy during delamination of the faceplates has been measured for mode I loading conditions. The flocked fiber core fails by pull-out of fibers from their sockets in the adhesive. A model has been developed to predict the energy absorbed during fiber pull-out, based on simple shear lag theory. Good agreement is found between theory and experiment, assuming an interfacial shear strength between fiber and adhesive of about 5 MPa. The long in-plane mesh and the short fiber 3-D array both fail by fracture within the core. A model has been developed for prediction of the fracture energy for this type of failure, based on deformation and fracture of individual fibers. Good agreement with experiment is obtained for the short fiber 3-D array sheet, which has the highest fracture energy. Experimental values for the long fiber in-plane mesh, however, are lower than predicted. This is ascribed to a tendency for delamination to occur by failure of solid-state sintered fiber-fiber necks, which are relatively weak, rather than by fiber fracture.

(d) Through-thickness electrical resistivities have been measured for the cores of the flocked sheet (~10 Ω.cm), adhesively bonded in-plane mesh (~1 Ω.cm), brazed long in-plane mesh (~0.01 Ω.cm), and the brazed short fiber 3-D array (~0.001 Ω.cm). These values compare well with predictions from simple geometrical models. The high resistivity of the flocked fiber core is attributed to poor electrical contact between the fiber ends and the faceplates.

(e) The flocked (adhesively bonded) material could not readily be welded. A shunt was required to achieve any melting, but electrical contact through the core was variable. In addition, weld flaws and faceplate damage occurred during welding. The brazed long fiber in-plane mesh material, on the other hand, exhibited good welding characteristics, with good current flow from the start of the welding period. The short fiber 3-D array sheet is also readily weldable.

(f) Overall, it is clear that sheet of this type can offer attractive combinations of lightweight, high beam stiffness/strength, good interfacial toughness, and low through-thickness electrical resistivity—allowing resistance welding to be carried out easily with conventional equipment. Of the three cores examined, the brazed short fiber 3-D array structure clearly offers the best combination of these properties. Ongoing work is aimed at exploring the influence of core structure on other aspects of sheet performance, including fatigue resistance, formability, energy absorption during crushing, and sound/vibration damping characteristics, as well as optimization of sheet manufacturing procedures. The material is evidently of potential interest to the automotive industry, even though it will inevitably be somewhat more expensive than monolithic metal sheet.

References

1. McCormack, T. M., R. Miller, O. Kesler, and L. J. Gibson, "Failure of Sandwich Beams with Metallic Foam Cores," *Int. J. of Solids and Structures*, Vol. 38 (28–29), 4901–4920, 2001.

2. Chen, C., A. M. Harte, and N. A. Fleck, "The Plastic Collapse of Sandwich Beams with a Metallic Foam Core," *Int. J. Mech. Sci.*, Vl. 43, No. 6, 1483–1506, 2001.

3. Harte, A. M., N. A. Fleck, and M. F. Ashby, "Sandwich Panel Design using Aluminum Alloy Foam," *Adv. Engng. Mater.*, vol. 2, No. 4, 219–222, 2000.

4. Lok, T. S., and Q. H. Chen, "Elastic Stiffness Properties and Behavior of Truss-Core Sandwich Panel," *J. Structural Engineering—ASCE*, Vol. 126, No. 5, 552–559, 2000.

5. Sypeck, D. J., and H. N. G. Wadley, "Multifunctional Microtruss Laminates: Textile Synthesis and Properties," *J. Mater. Res.*, Vol. 16, No. 3, 890–897, 2001.

6. Chiras, S., D. R. Mumm, A. G. Evans, N. Wicks, J. W. Hutchinson, K. Dharmasena, H. N. G. Wadley, and S. Fichter, "The Structural Performance of Near-Optimized Truss Core Panels," *Int. J. of Solids and Structures*, Vol. 39, No. 15, 4093–4115, 2002.

7. Gustavsson, R., "Formable Sandwich Construction Material and Use of the Material as Construction Material in Vehicles, Refrigerators, Boats etc.," patent WO 98/01295, 15th Jan., 1998, AB Volvo, International.

8. Markaki, A. E., and T. W. Clyne, "Mechanics of Thin Ultra-Light Stainless Steel Sandwich Sheet Material: Part I—Stiffness," *Acta Mater.*, Vol. 51, No. 5, 1341–1350, 2003.

9. Markaki, A. E., and T. W. Clyne, "Mechanics of Thin Ultra-Light Stainless Steel Sandwich Sheet Material: Part II—Resistance to Delamination," *Acta Mater.*, Vol. 51, No. 5, 1351–1357, 2003.

10. Jou, M., "Experimental investigation of resistance spot welding for sheet metals used in automotive industry," *JSME Int. J. Series C-Mech. Systems, Machine Elements, and Manufacturing*, Vol. 44, no. 2, 544–552, 2001.

11. Babu, S. S., M. L. Santella, Z. Feng, B. W. Riemer, and J. W. Cohron, "Empirical Model of Effects of Pressure and Temperature on Electrical Contact Resistance of Metals," *Science and Technology of Welding and Joining*, Vol. 6, no. 3, 126–132, 2001.

12. Wang, S. C., and P. S. Wei, "Modeling Dynamic Electrical Resistance during Resistance Spot Welding," *J. Heat Transfer-Trans.—ASME*, Vol. 123, No. 3, 576–585, 2001.

13. Khan, J. A., L. J. Xu, Y. J. Chao, and K. Broach, "Numerical Simulation of Resistance Spot Welding Process," *Numerical Heat Transfer Part A—Applications*, Vol. 37, No. 5, 425–446, 2000.

14. Dilthey, U., H. C. Bohlmann, U. Reisgen, W. Sudnik, W. Erofeew, and R. Kudinow, "Modelling and Numerical Simulation of Resistance Spot Welding with Experimental Verification," *9th International Conference on the Joining of Materials*, Helsingor, Denmark, 38–43, 1999.

15. Tang, H., W. Hou, S. J. Hu, and H. Zhang, "Force Characteristics of Resistance Spot Welding of Steels," *Welding Journal*, Vol. 79, No. 7, 175S–183S, 2000.

16. Zhou, Y., P. Gorman, W. Tan, and K. J. Ely, "Weldability of Thin Sheet Metals during Small-Scale Resistance Spot Welding using an Alternating-Current Power Supply," *J. Electronic Materials*, Vol. 29, No. 9, 1090–1099, 2000.

17. Markaki, A. E., S. A. Westgate, and T. W. Clyne, "The Stiffness and Weldability of an Ultra-Light Steel Sandwich Sheet Material with a Fibrous Metal Core," *Processing and Properties of Lightweight Cellular Metals and Structures*, Seattle, eds. A. K. Ghosh, T. D. Claar, and T. H. Sanders, TMS, 15–24, 2002.

18. Oberle, H., C. Commaret, R. Magnaud, C. Minier, and G. Pradere, "Optimizing Resistance Spot Welding Parameters for Vibration Damping Steel Sheets," *Welding Journal*, Vol. 77, No. 1, 8S–13S, 1998.

19. Sorensen, B. F., A. Horsewell, O. Jorgensen, and A. N. Kumar, "Fracture Resistance Measurement Method for *in situ* Observation of Crack Mechanisms," *J. Am. Ceram. Soc.*, vol. 81, no. 3, 661–669, 1998.

20. Whitehouse, A. F., C. M. Warwick, and T. W. Clyne, "The Electrical Resistivity of Copper Reinforced with Short Carbon Fibers," *J. Mat. Sci.*, Vol. 26, 6176–6182, 1991.

21. Underwood, E. E., *Quantitative Stereology*, Reading: Addison–Wesley Publishing Company, 1970.

22. Hull, D., and T. W. Clyne, *An Introduction to Composite Materials*, Cambridge Solid State Science Series, ed., D. R. Clarke, S. Suresh, and I. M. Ward, Cambridge: Cambridge University Press, 1996.

23. DiFrancia, C., T. C. Ward, and R. O. Claus, "The Single-fiber Pull-out Test 1: Review and Interpretation," *Composites A*, Vol. 27, 597–612, 1996.

24. Kim, J. K., C. Baillie, and Y. W. Mai, "Interfacial Debonding and Fiber Pull-out Stresses, Part I—Critical Comparison of Existing Theories with Experiments," *J. Mat. Sci.*, Vol. 27, 3143–3154, 1992.

Section 4

Processing and Manufacturing

The cost-effective scaling-up of new processes and new manufacturing techniques for the mass production of automobiles is a huge technological and economic challenge. The supply chain in the mass production automotive sector is notoriously competitive and production margins are often small. Even in niche automotive manufacture, there are significant constraints and downward pressure on costs, so that new materials and their associated fabrication and assembly procedures must offer demonstrable economic benefit. In recent years in some automotive segments, cost of ownership and especially fuel costs, have increased in importance and are beginning to favor the adoption of some new manufacturing processes—for example the hydroformed Al alloy sub-chassis—allowing more durable, more cost-effective and low life-cost materials to be used. This section discusses the drivers and opportunities for new processes and manufacturing technology including:

- Welding and joining
- Titanium alloys in harsh environments
- Casting
- Durable and high-performance composites
- Surface treatments in autosport

15

Welding and Joining

J. G. Wylde and J. M. Kell

CONTENTS

Introduction

Within the automotive sector there is a continual drive toward reducing costs, improving performance, and increasing sustainability. Inevitably, this leads to the search for new materials and structures that will offer improved performance and reduce cost. These efforts have resulted in developments and advances in materials across a spectrum of materials including metals, plastics, ceramics, and composites.

However, all too often, one vital ingredient is ignored. This is the simple question, how will these materials be joined? Thus, with few exceptions, new materials can only be effectively used in engineering structures if they can be joined to themselves, and in many cases, to other materials. These joins must be capable of being made cheaply and reliably in a mass production environment, and furthermore, the properties of the joints must be sufficient to avoid premature failure in service.

Consequently, materials joining technology is one of the key enabling technologies in almost every branch of manufacturing, and the automotive sector is no exception. Thus, automotive engineers are keenly interested in developments in joining technologies. Reducing the cost of manufacturing can have a major impact on price. Increasing the strength and integrity of joints can have a major impact on design, and offer the potential for reducing material thickness, and thus, reducing weight.

Dozens of different joining technologies are used in the manufacture of automotive components and structures. This chapter covers some of the recent developments in joining technology that provide opportunities for improved use of advanced materials, and improved aesthetic appearance and design.

Friction Stir Welding

In recent years, there has been an increasing interest in the use of lightweight materials in automotive fabrication. Aluminum alloys offer considerable potential for weight reduction because of their high strength-to-weight ratio. However, they are not generally as readily weldable, particularly in a mass production environment, as many steels. For this reason, friction stir welding (FSW) has gained increasing interest since its invention some ten years ago. Friction stir welding is a novel joining process developed at TWI in 1991.[1] Engineers have long recognized that frictional heating could be used to join materials. Conventionally, one round or tubular component is rotated, and pushed against another. The frictional heating that takes place causes both components to become hot, and one or both to become plasticized. The application of a forging force to push the components together can then be used to form a solid state or friction weld. Such welds have consistently been shown to possess exceptional mechanical properties.

During the past thirty years or so, a number of developments of the friction welding process have been made to allow the process to be applied to a wider range of geometries and shapes. Orbital and linear friction welding enabled the process to be applied to a variety of non-round components. However, virtually all of these techniques involve relative motion between parts to generate the frictional heating. This naturally limits the application of friction welding to relatively modest sections and components, which can be held within a machine and moved relative to each other to develop the frictional heating. Furthermore, they involve the joining process taking place at, more or less, the same time across the entire joint area. Thus, as the size or length of the component grows, so the forces involved in the application of the process also increase dramatically.

In the 1980s, Thomas[2] developed other variants of frictional heating of materials and showed that friction could also be used to extrude metals—friction extrusion. It was from this development and the desire to extend friction techniques to larger parts, that friction stir welding was born. Thomas et al.[3] discovered that butt-welded seams could be produced using a rotating tool

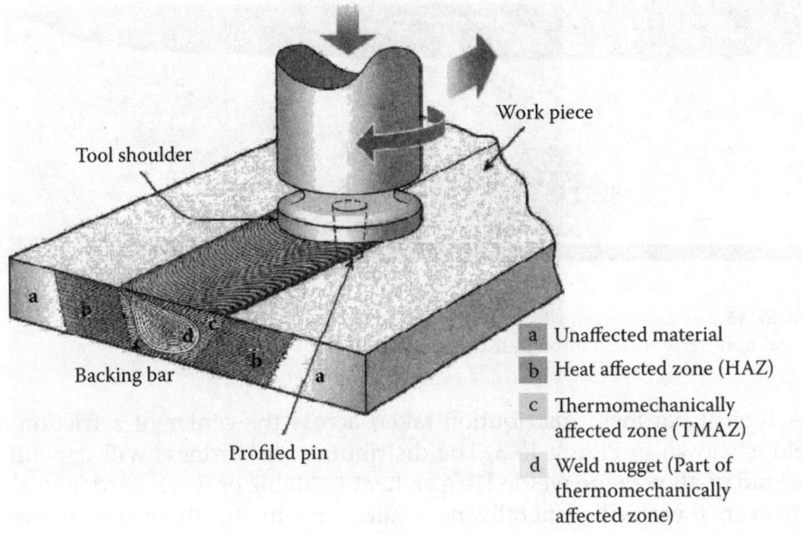

FIGURE 15.1
Principle of FSW.

with a shoulder and pin. The design of the tool is key to the successful application of the process, and a number of different designs have been investigated. Essentially, the height of the pin is the same as the thickness of the material to be joined. The pin is pushed into the seam until the shoulder comes into contact with the top surface of the material. After a brief dwell period, the tool is then moved along the length of the seam.

As the tool moves through the material, some material is taken from the edges of the parts being joined, mixed, and transported to the back of the pin where it extrudes into the area behind the pin to form a solid phase bond. This concept is illustrated in Figure 15.1. Being a friction process, there is no melting of the material, thus the weld produced is a solid phase joint. Consequently, there is no fume and no associated loss of alloying elements. There is no porosity, as there has been no solidification from molten material.

Figure 15.2 shows a transverse section through a typical friction stir weld. Threadgill[4] has characterized the various regions of the weld as indicated in the figure. The weld "nugget" shows a fully recrystallized material with a fine grain structure. Next to the nugget lies a zone of material that has been subject to considerable mechanical and thermal effects, the thermomechanically affected zone. Here, the original structure has been distorted by the mechanical motion of the tool. Some recrystallization has taken place and the material has been subjected to a thermal cycle. Adjacent to this, is a thermally affected region that forms a transition to base material.

FIGURE 15.2
Typical transverse section through friction stir weld.

A typical hardness distribution taken across the center of a friction stir weld is shown in Figure 15.3. The distribution of hardness will depend on the kind of alloy being welded, e.g., a heat-treatable or work-hardened alloy. However, there will generally be a softening in the thermomechanically affected region, and a consequent reduction in strength for the heat treatable alloys. However, invariably, researchers have discovered that the mechanical properties of friction stir welds are at least equal to, and generally exceed those, of arc-welded joints. In Figure 15.3, 5083-0 refers to annealed alloy AA5083, whereas 5083-H321 is the same alloy after cold-working and various heat treatments.

Industrial applications of friction stir welding were reported within five years of the invention of the process, driven by some advantages of the process:

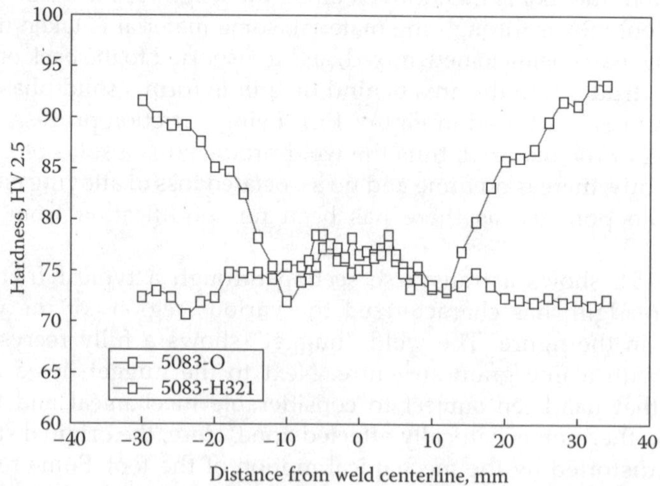

FIGURE 15.3
Typical hardness distribution through a friction weld.

- Very low distortion
- Fully mechanized process
- No fume, porosity, or spatter
- No melting of the base material
- Cost-effective for suitable applications

Although reported applications cover many industrial sectors, by far the most reported uses come from the broad area of transportation. Thus, applications have been reported for ships, railway vehicles, automotive components, space vehicles, and more recently, aircraft structures. New applications for the process are arriving on a regular basis.

In the automotive sector, current applications include wheel rims,[5] suspension arm struts,[6] and a variety of body components.[7,8] Figure 15.4 shows some typical examples.

In terms of the mechanical properties of friction stir welds, it is important to note that far more data are required before any definitive conclusions can be drawn concerning their long-term engineering performance. The two properties that are generally looked for by designers of aluminum structures are tensile properties, i.e., proof and tensile strength, and fatigue strength.

(a)

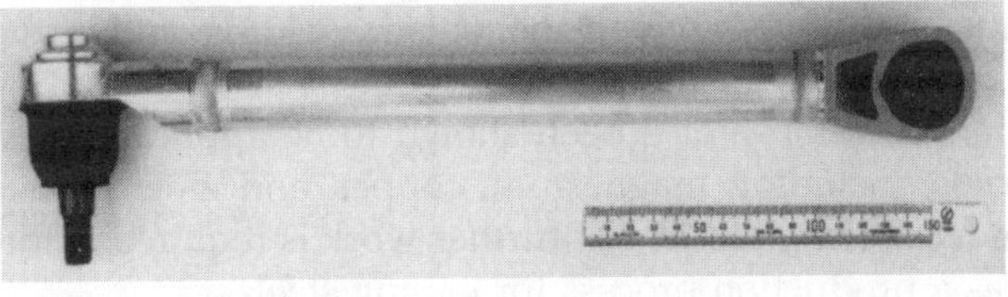

(b)

FIGURE 15.4
Examples of friction stir welding applications. (a) Wheel rim—photograph courtesy of Simmonds Wheels P/L. (b) Suspension arm strut—photograph courtesy of Showa Denko.

In terms of tensile strength, various researchers have carried out 180° face and root bend tests, and cross-weld tensile tests. The bend test is a useful guide to weld quality and ductility, and provided friction stir welds are made under optimum conditions, the bend test is invariably passed without evidence of cracking. In terms of the cross-weld tensile tests, many comparisons have been made with base material properties and with other welding methods. The results vary according to the type of alloy being tested, but are generally very encouraging.

In terms of fatigue performance, friction stir welds demonstrate good properties. In general terms, welded joints possess a much lower fatigue strength than that of the base material. There is a combination of reasons for this including the presence of sharp discontinuities on a microscopic scale at the edge of the weld, a stress concentration caused by the weld shape, and the presence of tensile residual stresses. This phenomenon is well understood by designers, and various sets of fatigue design curves have been developed for arc-welded joints. Thus, the design stress for a conventional-welded joint can be determined by establishing the appropriate joint classification and determining the design stress according to the required fatigue life.

Remarkably, fatigue tests on friction stir welds have indicated that they can possess a higher fatigue strength than arc welds. In some cases there seems to be very little difference between the fatigue performance of the friction stir weld and that of the base material. The reasons for this are not fully understood, and more tests are required to verify this conclusion. Preliminary results suggest that the design stress for a friction stir weld might be some 50% higher than that for an arc weld. Clearly, if this increase is confirmed with additional data, it will have very significant consequences for designers of automotive structures.

In general, in fatigue-sensitive structures it is the fatigue strength of the welds that determines the design stresses, and hence, the material thicknesses. If it is confirmed that a higher design stress can be used for friction stir welding, then this may result in some structures being fabricated from thinner section materials, and this could have major significance for the automotive industry where the use of thinner materials will reduce weight, increase performance characteristics, and improve fuel economy. Consequently, it is not surprising that many designers and fabricators are investigating the potential for friction stir welding.

The vast majority of research on friction stir welding, and virtually all the production applications of the process to date, relate to aluminum alloys. However, the process has also been shown to be applicable to a range of other materials, including magnesium, copper, and zinc. It has also been shown to be feasible for steels, but further work is required before it can be considered as a production process for joining steels.

One final and particularly interesting feature of friction stir welding is its ability to make joints between some dissimilar materials. For example, joints between different aluminum alloys are feasible, as are joints between cast and wrought aluminum alloys. This flexibility is particularly relevant to the

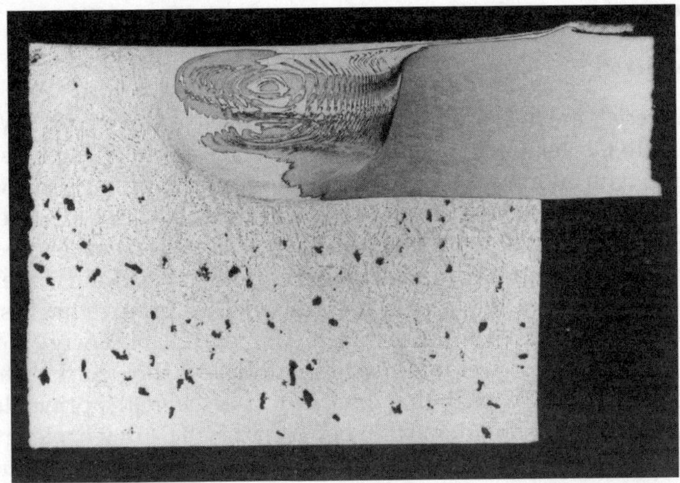

FIGURE 15.5
Wrought aluminum welded to cast aluminum alloy.

automotive sector, where there is an increasing interest in the use of cast components. Figure 15.5 shows an example of a friction stir weld between cast and wrought aluminum alloys.

Friction stir welding has also been used with some success to join completely dissimilar materials, e.g., aluminum and magnesium alloys. This is particularly relevant to the automotive sector as there is increasing interest in the use of magnesium to reduce weight. Figure 15.6 shows a joint between aluminum alloy 2219 and magnesium alloy AZ 91.

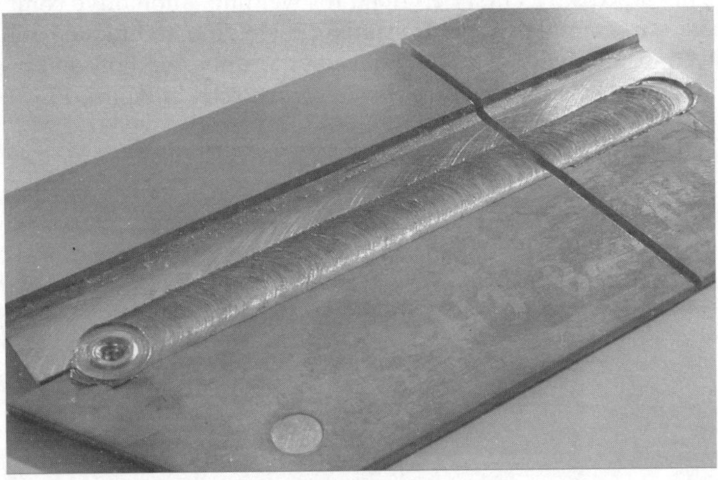

FIGURE 15.6
Friction stir weld between aluminum and magnesium alloys.

Clearweld™

Plastic materials are increasingly used in automotive components. TWI in conjunction with Gentex Corporation have developed a technique for laser welding plastics with an infrared absorbing material, creating a joint that is almost invisible to the human eye. Typically, carbon black would be used as the absorbing medium for the laser light; although the new approach enables two similar clear (or colored) plastics to be joined, with a minimal marking at weld line.

The development of the laser as an industrial heat source has resulted in a range of applications that utilize the precise, controllable energy it delivers. Early developments in welding plastics with lasers showed that thin films could be joined.[9] However, at that time, CO_2 lasers were the principle power source. The nature of the interaction between the 10.6 µm wavelength beam from the CO_2 laser and thermoplastic materials meant an analogue to the deep-penetration process used to weld metals could not be developed. The CO_2 laser beam is absorbed at the surface of the plastic, relying on conduction of heat through the thickness of the material, which results in decomposition, vaporization, and charring, before any significant depth of material is melted. Nonetheless, thin polyolefin films, of the order 0.1 mm thick, have been successfully welded with a CO_2 laser at speeds up to 500 m/min.[10]

The increasing use of Nd:YAG solid-state lasers, and the advent of diode lasers (both producing beams with a near infrared wavelength), has made available lasers with different beam/material interaction characteristics with thermoplastics. In transmission welding, the laser beam passes through the top (transmitting) layer and is absorbed by the filler in the lower layer, producing sufficient heat to make a weld at the interface between the two parts. This process was first described for welding automotive components in 1985.[11] An example of the transmission-welding technique, utilizing a visually transmissive plastic material for the upper section and a carbon black loaded plastic for the lower layer, can be seen in Figure 15.7.

FIGURE 15.7
Laser transmission weld in 4 mm thick polypropylene using a 100W Nd:YAG laser at a speed of 1.6 m/min. The weld is at the interface between the light and dark materials.

FIGURE 15.8
Laser overlap weld in clear 3 mm thick PMMA made with an infrared dye impregnated film at the interface.

An extension of the transmission laser-welding process that allows completely clear or similarly colored components to be welded has also been described.[12] This technique uses an infrared absorbing material, clear in the visible range of the spectrum, but tailored to absorb strongly the specific wavelength of the laser beam at the interface of the materials to be joined. Infrared absorbing pigments are also available as an alternative. The nature of the absorbing material means the laser wavelength is absorbed with high efficiency, thus requiring relatively small amounts at the interface between the two components to be joined. Initial development work on the process was mainly carried out using polymethylmethacrylate (PMMA) test specimens, and an example of an overlap weld made by applying a painted layer of absorbent material at the joint region between two transparent sheets of 3 mm thick polymethyl methacrylate is shown in Figure 15.8.

Although the example in Figure 15.8 is shown with two visibly clear sheets of polymethyl methacrylate, the process can be used to join several other materials, colored or otherwise.

The absorbing material at the interface between the materials acts as the site where the light from the laser is absorbed and converted into heat in a well-defined area. The area of heating, and hence joining, may be defined by either the size of the laser beam, or the coverage of the absorbing material used. In the experiments reported here, both Nd:YAG and diode laser light have been used. Both these laser wavelengths are easily transmitted via optical fibers, which enhances the flexibility of the process in industrial terms.

The process has also been applied to fabrics. Nd:YAG lasers are usually employed in a de-focus position to produce a spot of laser energy some few mm in diameter. This energy profile is almost ideal for the fabric-welding process. The welding occurs as the heat generated in the absorbing material is sufficient to melt, of the order ≈ 0.1 mm of the polymer fabric. The heat

FIGURE 15.9
Continuous overlap welds made using infrared absorbing dye in the fabric Goretex™.

generation at the interface is controlled by the absorption coefficient of the layer and the processing parameters. The main parameters for a given width of weld are laser power, energy distribution in the focus, and the welding speed.

For these experiments, an Nd:YAG laser with a 7 mm diameter focal spot was used at powers between 50 and 100 watts, and welding speeds in the range 500–1000 mm/min.

Figure 15.9 shows continuous, hermetic overlap welds made in the water-proof fabric Goretex™ using an Nd:YAG laser beam of approximately 100 W in power, and a welding speed of 500 mm/min.

Peel and lap/shear tests were performed on 25 mm wide samples of joined material at a test rate of 5 mm/min, and the results are shown in Table 15.1. The test results are quoted as the maximum applied force per mm of seam. As a percentage of the strength of the parent materials, 25% to 40% strengths were obtained for the welded joints in a simple lap configuration.

The work has shown that polymer fabric materials can now be laser welded using near infrared absorbing material as a mechanism to produce heat and localized melting. The welds produced are cosmetically appealing, and the upper and lower surfaces of the material are unaffected by the process.

TABLE 15.1

Results of Mechanical Testing on a Range of Woven Fabrics

Material Color	Thickness (mm)	Peel Strength (N/mm)	Lap/Shear Strength (N/mm)	Parent Strength (N/mm)
Brown	0.19	0.70	2.08	8.47
Orange	0.23	2.16	5.22	13.95
Bronze	0.16	2.07	2.76	9.38
Yellow	0.41	4.40	6.79	16.12

In mechanical testing, joint strengths of between 20% and 40% of the parent material strength have been achieved in a simple lap joint. The welding process is efficiently achieved using compact diode laser sources now available commercially, and lends itself easily to high levels of automation.

The process of laser welding using an infrared absorbing material has been given the trademark ClearWeld™. In addition, patent protection has been initiated by TWI on this process. Gentex Corporation is licensed by TWI to exploit the ClearWeld™ technique.

AdhFAST™

Despite widespread skepticism, adhesives are playing an increasingly important role throughout the engineering world, and will continue to find new applications in volume applications. The primary drivers behind the growth in the use of adhesives are the increasing interest in combining different materials in structures to maximize performance, e.g., plastics, metals, ceramics, composites, etc., and the advantages adhesives offer over traditional point joining techniques. These advantages can be broadly defined as:

- Ability to join almost any material combination.
- Superior fatigue properties.
- Elimination/reduction of stress concentration points by bonding the whole joint area.
- Ability to have mechanical properties tailored to joint function, i.e., rigidity, elasticity, toughness, coefficient of thermal expansion, etc.
- Ability to have physical properties tailored to requirements, i.e., electrical and thermal conduction/insulation, and cure initiated by radiation (blue or UV light, electron beams).
- Sealing ability.
- Elimination/minimization of thermal distortion.

However, despite these advantages, some engineers are concerned about the use of adhesives because of the perception that they are:

- A poor or weaker substitute for welding or mechanical fasteners.
- Messy to use.
- Perceived to have significant health and safety risks.
- Difficult to inspect and to assess the significance of any defects.
- In need of complex pre- and post-processing.

It is well established that adhesives can fulfill a structural function both reliably and effectively. For example, the brake shoes in most cars are bonded;

aircraft rely upon adhesive in conjunction with rivets to bond and seal the fuselage; and composite-bonded drive shafts are used in lorries and cars. Although it is often true that an adhesive bond may never be as strong as the parent material, this can frequently be compensated for in the design of the joint to significantly enhance the overall performance of the structure. It is also accepted that the quality control (QC) behind such bonding operations can be difficult and complex, especially when the structural integrity of the joint is brought into question. These concerns necessitate the need for skilled personnel in both the design and the implementation stages of production. Without the appropriate skills, quality control confidence will be limited and there is a greater chance of failure, either though poor joint design/material selection, or during the assembly stages.

Adhesives that cure at room temperature usually consist of two components that need to be mixed prior to application, and the process of applying the adhesive to the surfaces of the substrates can be messy and time consuming. For many structures, it is common, and indeed best practice, to dry-assemble the joint first, to check for tolerance. The adhesive is then applied before reassembly in conjunction with jigging, which is used to hold the joint together while the adhesive cures and hardens. In basic terms, the process can be defined as follows:

> **Surface preparation on the materials to be joined.** This may be a simple degrease operation, but usually more complex processes are employed. For example, abrasion (hand papers, grit blast, or shot blast), chemical etching, anodizing, priming, use of coupling agents, flame plasma, or corona discharge are all used.

> **Mixing and application of the adhesive.** This is often a manual operation where beads of adhesive are applied to one or both of the surfaces with an adhesive dispenser combined with a mixing unit. To ensure complete wetting of the surfaces to be bonded, the adhesive can then be spread out evenly over the surfaces.

> **Assembly of components.** The components then have to be assembled and aligned correctly. This process can be messy if excess adhesive has been used. It is also often difficult to achieve accurate alignment of the components without special jigging and guides.

> **Additional jigging and clamping.** Such jigging/clamping is often required once the structure has been assembled, to apply an even pressure in the joint area while the adhesive cures.

> **Curing of the adhesive.** Many adhesives have been formulated to cure at room temperature through reactive chemistry, but there are others that require heat to react, therefore necessitating the use of an oven or heating equipment.

> **Disassembly and checking of the structure.** Once cured, time has to be spent removing the structure from the jigging, and checking that the adhesive has fully filled the joints (visual inspection).

For an adhesively bonded joint to be reliable, the bonding process must follow strict procedures by trained operators. Many companies do not appreciate the need for skilled staff, and this can result in failure. The arising unreliability can reinforce the perception that the adhesive is at fault, rather than the process.

One way in which such problems are overcome is by combining a mechanical fastening system such as riveting, bolting, clinching, or spot welding with the adhesive to form a *hybrid* joint. The point-fastening system enables a safety back-up to be built into the joint (and some visible confidence), while retaining the superiority of the adhesive joint, especially in terms of fatigue, and sealing capacity.

It is from this background that the concept of AdhFAST™ arose.[13] AdhFAST™ is a hybrid joining *system* that brings together the advantages of both adhesives and fasteners, and in addition, offers a high degree of quality control with a minimum of additional operator training. In essence, AdhFAST™ takes the form of a four-function fastener that:

- Locates—enables positional accuracy between components to be defined.

- Fastens—traditional function, plus acts as a jigging aid during the adhesive cure stage.

- Spaces—controls the spacing between the materials to be joined, thereby enabling adhesive to be easily injected, and defining the final thickness of adhesive in the joint.

- Allows adhesive to be injected—accomplished either though a central hole or down features on the sides of the fastener.

The fastener, which can take a range of forms (nut and bolt, screw, rivet, etc.), fulfills its function as a fastener in that it locates and fastens the materials to be bonded together. The fastener is positioned such that it sits within the prospective joint away from edges and high-stress areas. In addition to its normal function, the fastener contains a spacer element (a shaped washer or similar), which contains grooves or features that will allow a gap to be maintained between the two spacers. In turn, the fastener is designed in such a way as to allow liquid or paste adhesive to be injected through or past it, and around the spacer element into the bond cavity. The adhesive can, therefore, be pumped into the bondline to fully fill the joint from the inside out. Provided appropriate surface preparation has been carried out on both surfaces to be bonded, the joint will be fully wetted by the adhesive prior to curing.

By using fasteners, the assembly and bonding process is simplified:

Surface preparation, the materials to be joined. This part of the operation cannot be changed as the type of pre-treatment defines the level of adhesion attainable to the surface of each substrate.

Assembly of components. The components are aligned, assembled, and held together using fasteners. The structure and associated joints can be quite complex in shape in that adhesive injection allows more than one component to be bonded at any one time. A structure using fasteners will not require additional external jigging. The edges of the joint may need to be sealed, which can be done in a number of ways including using adhesive release tape, using inflatable bellows, or a simple gasket.

Adhesive injection. The adhesive is then mixed as normal and injected into the joint cavity through the fasteners. As the adhesive fills the joint, its progress can be monitored by its appearance out of the hole in the next fastener. The injection process is then continued through that fastener after sealing the previous one. The amount of adhesive that the operator is exposed to is minimal.

Curing of the adhesive. As described previously, the adhesive cures either on its own, or with the application of heat or some other energy source.

Disassembly and checking of the structure. The only disassembly needed may be the peeling away of sealing tape, as no additional jigging is required. Visual inspection is as usual.

Employing fasteners enables the following further benefits:

- No external jigging
- Simplified dry assembly with accurate location and checks of tolerance
- Protection of pre-treated surfaces prior to bonding from excessive atmospheric exposure and operator contamination
- Minimal operator exposure to uncured adhesive
- Simplified adhesive application process
- Accurate metering of adhesive within the joint
- Accurate bondline control
- Saving in time due to elimination of jigging assembly/disassembly

In addition to the above benefits, fasteners offer a change in the manufacturing approach to bonding by breaking the linearity of the process, i.e., there can now be a dry assembly stage followed by a separate injection stage. In reality, this means that these operations could be done in different geographical locations, or at different times, depending upon production and manpower resources. With the correct selection of surface pre-treatment where a bonding window of days or weeks was possible, the storage of dry assembled parts ready for bonding, with the possibility of disassembly and re-use should an order be changed or amended, is made possible.

FIGURE 15.10
Typical AdhFAST™ joint.

Industries likely to benefit from a hybrid fastener/adhesive approach include consumer goods, aerospace, automotive, railway, and shipbuilding. A typical AdhFAST™ joint is shown in Figure 15.10.

Laser Welding of High-Strength Steels

Despite the considerable increase in the use of aluminum and magnesium alloys and other advanced materials, it is almost certain that steels will be continued to be used for vehicle production for many years to come. High-strength steels (UTS > 600 N/mm^2) are increasingly used to meet the severe requirements imposed by the automotive industry in terms of safety, reliability, and reduction in gauge for energy saving.[14] TRIP (transformation induced plasticity) steels have become of considerable interest in recent years because of their exceptional combination of high strength and ductility. Resistance spot welding is the main joining method for these steels but other methods such as laser welding are increasingly being investigated in the automotive industry, where there remains a need for weldability and joint performance data for these steels.

The high carbon equivalent in TRIP steels, coupled with fast weld-cooling rates associated with the welding process, leads to high hardness levels (up to 580 HV) in the weldment. Typically, resistance spot welding is achieved using conventional procedures, which give interface (brittle) fracture on testing, and make the welds unable to meet current automotive welding standards. The restricted performance is linked to the high hardness/low toughness levels within the weld nugget and heat-affected zone (HAZ). Modified welding procedures, such as long weld times and post-weld tempering,[15] have been suggested to reduce this weld brittleness. However, these are not always feasible or practical because of the increased cycle time, and their effects on the static and dynamic properties are not yet clear. Given the continued move toward the implementation of higher strength steels in the automotive industry, the benefit of achieving plug weld fracture modes associated with high-quality welds for TRIP steels is clearly evident.

Laser welding is increasingly used in the automotive industry as an alternative to resistance spot welding. It is generally recognized that a continuously welded joint can provide increased stiffness compared to an equivalent resistance spot-welded joint.[16] Further development of laser welding techniques, such as twin-spot beam and laser-arc hybrid, also has the potential to reduce the susceptibility to cracking,[17] and to maximize the benefits of TRIP steels. There is little information presently available related to the laser weldability and weld performance of TRIP steels.

This serves to demonstrate the importance of information relating to materials weldability before any material can successfully be used in production. TWI and others are researching many of these issues.

Conclusions

The strive to reduce cost, and improve performance and sustainability continue to interest automotive engineers in the new materials. However, all too frequently, engineers ignore the question of how such materials can be joined until far too late in the design and manufacturing process. Thus, the ability to weld or join materials safely and cost effectively in a production environment is vital to the successful application of new materials to engineering structures.

Advances in materials joining technology continue to meet the challenges provided by new materials development, and offer new opportunities for designers and manufacturers of automotive products. This chapter has briefly looked at a number of recent developments in joining technology that offer potential for joining a number of similar and dissimilar material combinations, and introduced some of the areas in joining technology that will receive the attention of the automotive industry in coming years.

References

1. Thomas, W. M., Nicholas, E. D., Needham, J. C., Murch, M. G., Temple-Smith, P., and Dawes, C. J., *Improvements relating to friction welding*, International patent application PCT/GB92, 6 December 1991.
2. Thomas, W. M., "Leading edge–Friction extrusion of powder and machine swarf," *TWI Connect*, July 1992.
3. Dawes, C. J., and Thomas, W. M., "Friction stir joining of aluminum alloys," *TWI Bulletin*, November/December 1995.
4. Threadgill, P. L., "Friction stir welding—the state of the art," *TWI Members Report* No. 678/1999, May 1999.
5. Simmonds, T., "Friction stir welding and alloy car wheel manufacture," *3rd International FSW symposium*, 27–28 September 2001, Kobe, Japan.
6. Sato, S., M. Enomoto, R. Kato, and K. Uchino, "Application of friction stir welding to suspension arms," *IBEC 98 International Body Engineering Conference & Exhibition*, Detroit, MI, 29 Sep–1 Oct 1998.
7. Smith, C. B., "Robotic friction stir welding using a standard industrial robot," *2nd International Symposium on FSW*, 26–28 June 2000, Gothenburg, Sweden.
8. Kallee, S., and A. Mistry, "Friction stir welding in the automotive body in white production," *1st International Symposium on FSW*, Thousand Oaks, CA, 14–16 June 1999.
9. Silvers, H. J., Jr., and S. Wachtell, "Perforating, welding and cutting plastic films with a continuous CO_2 laser," Pennsylvania State University, *Eng. Proc.*, 88–97, August 1970.
10. Jones, I. A., and N. S. Taylor, "High speed welding of plastics using lasers," *ANTEC '94 conference proceedings*, 1–5 May 1994, San Francisco, CA.
11. Jidosha, K. K., "Laser beam welding of plastic plates," Patent Application JP85213304, 26 September 1985.
12. Jones, I. A., P. A. Hilton, R. Salavanti, and J. Griffiths, "Use of Infrared Dyes for Transmission Laser Welding of Plastics," *ICALEO*.
13. Wylde, J. G., C. S. Punshon, G. C. McGrath, P. M. Burling, E. J. C. Kellar, A. Taylor, and I. A. Jones, "Recent Innovations in Materials and Joining at TWI," *New-Wave of Welding and Joining Research for the 21st Century Conference*, March 22–23, 2001, Osaka, Japan.
14. Drewes, E., B. Engl, and U. Tenhaven, "Potential for lightweight car body construction using steel," *Technische Mitteilungen*, 1994 (1) 25–32.
15. Chuko, W., and J. E. Gould, "Development of appropriate resistance spot welding practice for transformation-hardened steels," *Sheet Metal Welding Conference IX*, October 17–20, 2000.
16. Irving, B., "Building tomorrow's automobiles," *Welding Journal*, Vol. 74, no. 8, 28–34, 1995.
17. Xie, J., "Dual-beam laser welding and its applications," *Sheet Metal Welding Conference IX*, October 17–20, 2000.

16

Titanium Aluminide-Based Intermetallic Alloys

Takayuki Takasugi

CONTENTS

Introduction

Intermetallic alloys based on gamma (γ) TiAl are a potentially important vehicle and aerospace structural materials because of their light weight, relatively good high-temperature mechanical properties, and oxidation resistance.[1-6] One benefit of intermetallic alloys based on γ-TiAl is the variety of microstructures that can be contrived, such as a γ grain microstructure, a duplex microstructure consisting of γ and γ/α_2, a dual phase micro- structure consisting of distinct γ and γ_2 grains, and a fully γ/α_2 lamellar microstructure by alloy modifications and microstructural control. However, many of the microstructures typically exhibit low ductility and fracture toughness at ambient temperature. This low ductility at ambient temperature has widely been attributed not only to *intrinsic* factors, but also to *extrinsic* factors such as the so-called environmental embrittlement that can occur in intermetallic alloys based on γ-TiAl.[7-10] In these cases, hydrogen is introduced into the alloy microstructure from test atmospheres where moisture has been suggested to be able to react with the alloy, and to generate atomic hydrogen, resulting in reduced tensile elongation in air. However, it has not yet been known how the moisture-induced embrittlement is affected by microstructure, or by constituent phases in intermetallic alloys based on γ-TiAl. This chapter discusses TiAl-based intermetallic alloys of varying compositions made by isothermal forging and heat treatment to develop various kinds of microstructures. These alloys, together with lower cost Ti feedstock under investigation elsewhere, could allow TiAl-based alloys to find much wider application in the automotive sector.

TiAl-Based Alloys

Four kinds of TiAl-based intermetallic alloys denoted Alloys I, II, III, and IV, respectively, are discussed in this chapter. The chemical composition and fabrication procedure of the alloys are shown in Table 16.1. All the alloys were isothermally forged and alloys I and II were hot isostatically pressed (HIPed) at 1473 K for 2 h after isothermal forging.

To obtain the various microstructures, specimens were annealed in vacuum at high temperature and held for a predetermined time, followed by cooling to room temperature at a controlled cooling rate. Tensile specimens were prepared by electro discharge machining (EDM) from the isothermally forged TiAl-based ingots. Tensile tests were conducted at a fixed strain rate of 1.67×10^{-5} or 1.67×10^{-6} s^{-1} in air and vacuum as a function of temperature. The fracture stress (or tensile elongation) measured in air generally begins to increase rapidly at a certain temperature, characterized as the brittle-ductile transition temperature (BDTT). The degree of any moisture-induced embrittlement of the alloy was evaluated from the brittle-ductile transition temperature, at which tensile elongation in air becomes almost identical to that in vacuum. The fracture surfaces of deformed specimens were also examined by scanning electron microscopy (SEM).

Metallographic, chemical and structural observations of TiAl-based intermetallic alloys were carried out by optical microscopy (OM), x-ray diffraction (XRD), scanning electron micrograph with attached wavelength dispersive spectroscopy (WDS), and electron backscattering pattern (EBSP) analysis. Microstructural parameters, such as grain size or lamellar spacing in fully lamellar microstructure, were quantified on the basis of optical and scanning electron microscope observation. When calculating lamellar spacing in the fully lamellar microstructures, the difference among the three types of interfaces, i.e., γ/γ, γ/α_2 and α_2/α_2 was not taken into consideration. The determination of the constituent phases and their volume fraction was based on the results from WDS and EBSP.

TABLE 16.1

Alloy Compositions, Fabrication Procedures, and Resulting Microstructures of TiAl-Based Intermetallic Alloys

Alloy	Alloy Composition (at%)	Fabrication	Prepared (or Received) Microstructures	Grain Size (μm)	Lamellar Spacing (μm)	Volume Fraction of α_2 Phase (%)
I	Ti-46Al-7Nb-1.5Cr	Skull-melting in induction furnace Annealing at 1473 K for 5 h Isothermal forging at 1308 K to one third reduction HIP at 1473 K for 2 h under 200 MPa	γ (as-received)	20	—	0.2
			Dual phase	40	—	9
			Duplex	25	—	5
			Fully lamellar	100	0.5	26
				100	0.2	28
II	Ti-48Al-2Nb-2Cr	Skull-melting in induction furnace Annealing at 1473 K for 5 h Isothermal forging at 1308 K to one third reduction HIP at 1473 K for 2 h under 200 MPa	γ (as-received)	50	—	0.5
			Dual phase	45	—	10
III	Ti-48Al-2Mo	Arc-melting Isothermal forging at 1573 K	Duplex (as-received)	30	—	4
IV	Ti-52Al	Cold crucible induction melting (CCIM) Isothermal forging at 1373 K to one third reduction	γ (as-received)	50	—	0

Moisture-Induced Embrittlement

For alloy I, four kinds of microstructures were prepared after heat treatment, as shown in Figure 16.1. Figure 16.1 shows (a) a γ grain microstructure, (b) a duplex microstructure consisting of γ and γ/α_2, (c) a dual phase microstructure consisting of γ and α_2, and (d) a fully lamellar microstructure consisting of γ/α_2. Regarding the fully lamellar microstructure, two lamellar spacings of 0.5 and 0.2 μm resulted from the two cooling rates of 5.34 and 13.36 K min^{-1} respectively, after annealing at 1603 K. Primary grain sizes in alloy I varied widely across the different microstructures, ~20 μm in dual phase microstructures to ~100 μm in fully lamellar microstructures. The volume fractions of α_2 phase contained in dual phase, duplex, and fully lamellar microstructures were approximately 9%, 5%, and 26% ~ 28%, respectively, as shown in Table 16.1. Even the specimen ascribed to having γ grain only microstructures contained a trace fraction of α_2 phase of typically ~0.2%.

For alloy II, γ and dual phase microstructures were prepared. Here, the volume fractions of α_2 phase contained in γ grain microstructure and dual phase microstructure were approximately 0.5% and 10%, respectively (Table 16.1). For the alloys III and IV, dual phase and γ grain microstructures only were prepared respectively, with ~4% of α_2 phase in alloy III.

Figure 16.2 shows the variation in fracture stress with temperature for alloy I, with (a) γ, (b) duplex, and (c) dual phase microstructures. The specimens were deformed at a strain rate of $1.67 \times 10^{-5}\,\text{s}^{-1}$ in air and vacuum. For all microstructures, fracture stresses in vacuum were consistently higher

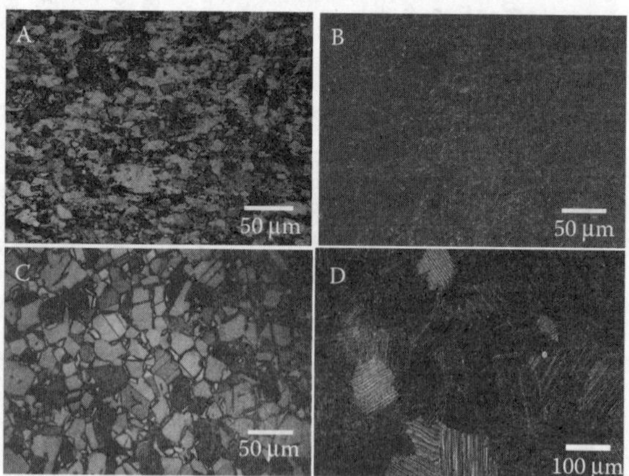

FIGURE 16.1

Optical microstructures of alloy I with (a) a γ grain microstructure, (b) a duplex microstructure consisting of γ and γ/α_2, (c) a dual phase microstructure consisting of γ and α_2, and (d) a fully lamellar microstructure consisting of γ/α_2.

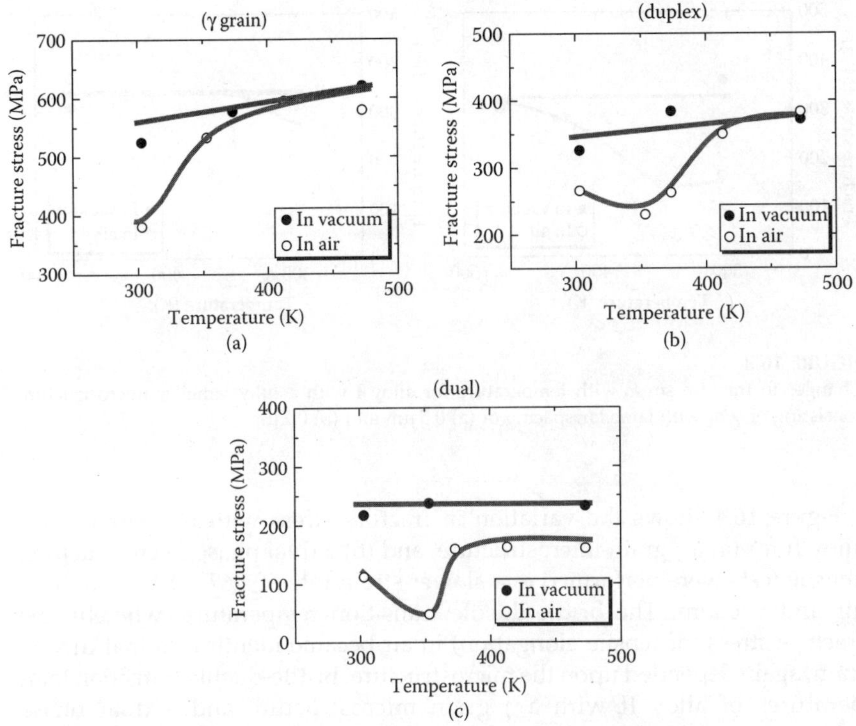

FIGURE 16.2

Changes in fracture stress with temperature for alloy I with (a) a γ grain microstructure, (b) duplex microstructure consisting of γ and γ/α_2, and (c) dual phase microstructure consisting of γ and α_2.

and almost insensitive to temperature. On the other hand, fracture stresses in air were lower than those in vacuum at low temperatures, but tended to recover to those in vacuum at high temperatures. Thus, brittle-ductile transition temperature, where fracture stress in air is defined to become identical to that in vacuum, depended upon microstructure. The brittle-ductile transition temperature of alloy I with a γ grain microstructure, duplex microstructure, and dual phase microstructure were approximately 353 K, 413 K and, > 473 K, respectively. Consequently, it is suggested that alloy I with γ grain microstructure is most resistant to moisture-induced embrittlement, and that alloy I with a dual phase microstructure is the most susceptible to any moisture-induced embrittlement. Regarding the effect of lamellar spacing in fully lamellar microstructure on the brittle-ductile transition temperature, Figure 16.3b shows that alloy I with a relatively fine lamellar spacing of 0.2 μm had a brittle-ductile transition temperature of 353 K, whereas a coarser lamellar spacing of 0.5 μm in Figure 16.3a had a brittle-ductile transition temperature of ~473 K.

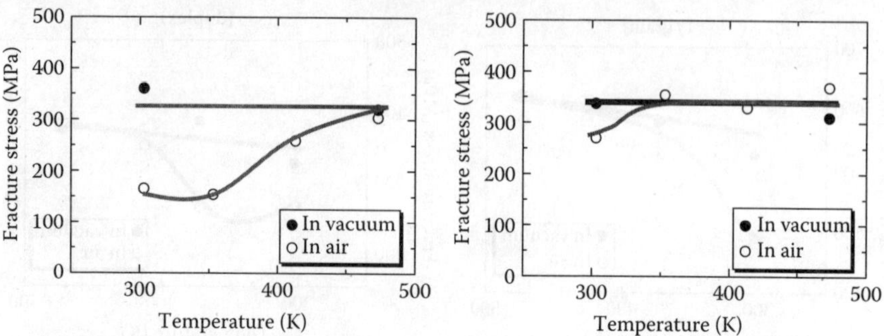

FIGURE 16.3

Changes in fracture stress with temperature for alloy I with a fully lamellar microstructure consisting of γ/α_2 with lamellar spacing of (a) 0.5 μm and (b) 0.2 μm.

Figure 16.4 shows the variation in fracture stress with temperature for alloy II for (a) a γ grain microstructure, and (b) a dual phase microstructure. Tensile tests were performed at a slower strain rate of $1.67 \times 10^{-6}\,\mathrm{s}^{-1}$ in both air and vacuum. The brittle-ductile transition temperature, whereby the fracture stress (or tensile elongation) in air became identical to that in vacuum, again depended upon the microstructure. Brittle-ductile transition temperatures of alloy II with a γ grain microstructure and a dual phase microstructure were approximately 413 K and >473 K respectively, indicating that alloy II with a γ grain microstructure was more resistant to the moisture-induced embrittlement than dual phase microstructures, and consistent with the behavior of alloy I.

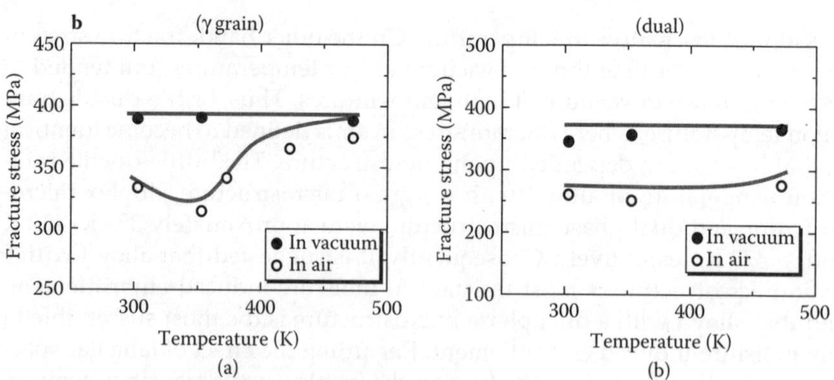

FIGURE 16.4

Variation in fracture stress with temperature for alloy II with a (a) γ grain microstructure, and (b) dual phase microstructure consisting of γ and α_2.

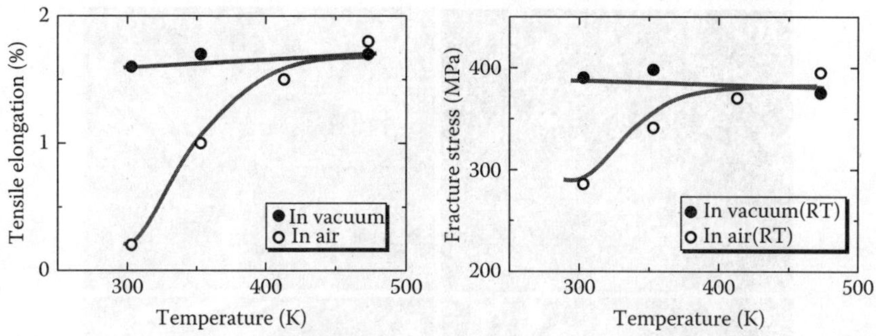

FIGURE 16.5

Variation of (a) tensile elongation and (b) fracture stress, with temperature for alloy III with duplex microstructure consisting of γ and γ/α_2.

Figure 16.5 shows the change in tensile elongation (or fracture stress) with temperature for alloy III with a duplex microstructure. Tensile tests were conducted at a strain rate of $1.67 \times 10^{-6}\,s^{-1}$ in air and vacuum. The brittle-ductile transition temperature was approximately 413 K. Interestingly, the determined brittle-ductile transition temperature was identical to that obtained for alloy I with the same duplex-type microstructure despite the differences in composition. Regarding alloy IV, tensile elongation as well as fracture stress was almost insensitive not only to testing atmosphere but also to strain rate, indicating that alloy IV with a γ grain microstructure was not susceptible to moisture-induced embrittlement at room temperature.

The appearance of the fracture surfaces of the TiAl-based intermetallic alloys was strongly dependent on microstructure, but independent of temperature and testing atmosphere, irrespective of the alloys examined. As an example, Figure 16.6 shows a scanning electron micrograph of alloy I with (a) γ grain microstructure, (b) duplex microstructure, (c) dual phase microstructure, and (d) fully lamellar microstructure. Figure 16.6a shows that the specimen with a γ grain microstructure exhibited river patterns characterization of transgranular cleavage fracture, mixed with a small amount of smooth facets, indicating intergranular fracture. The specimen with duplex microstructure in Figure 16.6b again exhibited a river pattern for transgranular cleavage fracture in the γ phase, but was accompanied by a mixture of interlamellar and translamellar fracture in the γ/α_2 regions. Figure 16.6c shows that the dual phase microstructure exhibited transgranular cleavage fracture, accompanied by regions of smooth facets, indicating interfacial fracture between α_2 grains and γ phase matrix. Lastly, Figure 16.6d shows that the specimen with a fully lamellar microstructure exhibited interlamellar and translamellar fracture patterns, the tendency for which was dependant on the angle between loading axis and lamellar planes.

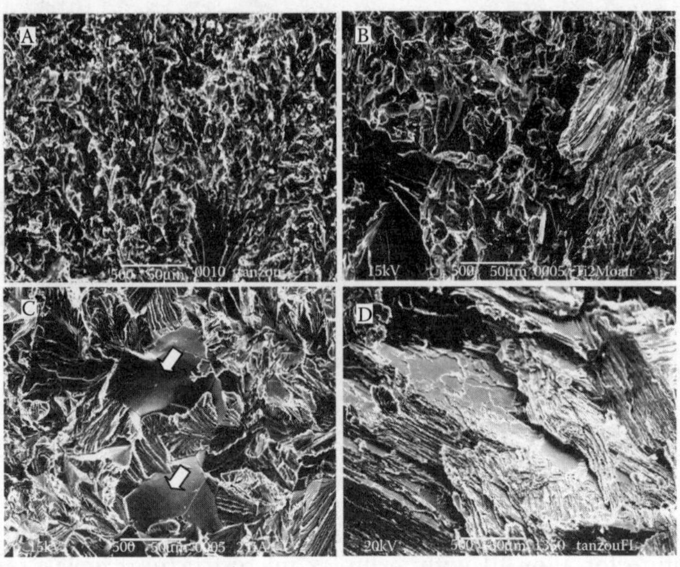

FIGURE 16.6
SEM fractographs of alloy I with a (a) γ grain microstructure, (b) duplex microstructure consisting of γ and γ/α_2, (c) dual phase microstructure consisting of γ and α_2, and (d) fully lamellar microstructure consisting of γ/α_2.

Embrittlement Mechanism

It has been suggested that the moisture-induced embrittlement of many intermetallic alloys including TiAl is induced at ambient temperature by decomposition of moisture on the alloy surface (or freshly exposed grain boundaries or cleavage planes), and subsequent micro-processes, such as permeation of atomic hydrogen into the alloy, and the migration and condensation of atomic hydrogen to grain boundaries (or lattice planes) in front of a propagating micro-crack. The grain boundary cohesion (or lattice cohesion) and the associated plastic work can be reduced by hydrogen condensation under the influence of a stress field ahead of a propagating micro-crack, thereby resulting in brittle intergranular fracture (or cleavage fracture).[11-13] It is suggested further that when the ambient temperature increases, the amount of absorbed hydrogen reduces, because of the lowering of the relative humidity, or from the formation of a protective surface oxide film. Consequently, hydrogen concentration does not reach the critical value to facilitate the subsequent propagation of a micro-crack. The recovery of tensile elongation occurs in a manner similar to the brittle-ductile transition with temperature, and the moisture-induced embrittlement generally disappears at high temperatures.[14,15]

Based on the measured temperatures at which the moisture-induced embrittlement of the TiAl-based intermetallic alloys disappears in this study, the γ grain microstructure is the most resistant to moisture-induced embrittlement, and the dual phase microstructure is the most susceptible to moisture-induced embrittlement. A duplex microstructure has moderate susceptibility/resistance to moisture-induced embrittlement. Similarly, a fully lamellar microstructure shows moderate susceptibility/resistance to moisture-induced embrittlement, and is dependent upon the lamellar spacing.

As the microstructural parameters relevant to the moisture-induced embrittlement of the TiAl-based intermetallic alloys, the volume fraction of α_2 phase, grain (or lamellar cell) size and grain (or lamellar) morphology must also be considered, and their values are shown in Table 16.1.

Let us firstly consider the moisture-induced embrittlement of the specimens with γ grain, dual phase, and duplex microstructures. Here, grain size may be excluded as a primary factor in determining moisture-induced embrittlement behavior, since alloy I with a duplex microstructure with the smallest grain size (25 μm) is not resistant to moisture-induced embrittlement; and alloy IV with a γ grain microstructure and the largest grain size (50 μm) is not the most susceptible to moisture-induced embrittlement (Table 16.1). However, moisture-induced embrittlement behavior becomes more pronounced as the volume fraction of α_2 phase increases. In other words, as the microstructure changes from a γ grain microstructure to a duplex microstructure through a dual phase microstructure, the moisture-induced embrittlement response becomes more severe. As shown in Table 16.1, the volume fraction of α_2 phase in the γ grain microstructure was 0.2, 0.5, and 0% in alloys I, II, and IV, respectively; 5 and 4% in duplex microstructure alloys I and III, respectively; and 9 and 10% in dual microstructure alloys I and II, respectively. Thus, the results shown here suggest strongly that α_2 volume fraction greatly affects the moisture-induced embrittlement of TiAl-based intermetallic alloys.

The precise role that α_2 phase plays in this embrittlement of TiAl-based alloys is unclear. One possible explanation is that the interface between α_2 phase grain and γ grain matrix acts as a preferable trap site for hydrogen atoms, and that interfacial cohesive strength is subsequently reduced, resulting in easier interfacial fracture. An alternative explanation may be that the α_2 phase grains scavenge atomic hydrogen from the γ grain matrix, but does not so effectively reduce the moisture-induced embrittlement of the alloy with this type of microstructure. The α_2 phase grains may be rather more susceptible to the moisture-induced embrittlement than other constituent phase (γ phase) by absorbing atomic hydrogen. It has been reported that Ti_3Al α_2-based alloys, such as Ti-25Al-10Nb-3V-1Mo and Ti-24Al-11Nb alloys, were embrittled by thermal exposure to hydrogen at high temperature, and their embrittlement was attributed to the formation of hydrides and their subsequent brittle fracture.[16-19]

The fully lamellar microstructure consisting of γ/α_2 showed moderate susceptibility to moisture-induced embrittlement, the extent of which strongly depended on the lamellar spacing, despite the large fraction of α_2 phase present. Because moisture-induced embrittlement becomes weaker as the lamellar spacing decreases, it is suggested that the morphology of the lamellae and the increase in the area of interface between γ and α_2 phases suppresses moisture-induced embrittlement. In contrast to the α_2/γ interfaces in the other microstructures, the α_2/γ interfaces in fully lamellar microstructures are composed of coherent (or semi-coherent) structures. Consequently, different hydrogen behavior and mechanical response may be expected at the α_2/γ interfaces in a fully lamellar microstructure. For example, the absorbed hydrogen may migrate and/or be trapped on the many interfaces between γ and α_2 lamellae, and consequently does not reach the critical concentration beyond which interlamellar fracture occurs. In other words, the large interfacial area between γ and α_2 lamellae and the high density of trap sites for hydrogen reduces hydrogen content on these planes, resulting in a reduction in moisture-induced embrittlement.

Summary

TiAl-based intermetallic alloys with various microstructures of γ grains, a duplex microstructure consisting of γ and γ/α_2, a dual phase microstructure consisting of γ and α_2, and a fully lamellar microstructure consisting of γ/α_2 were prepared from isothermally forged materials. These TiAl-based inter-metallic alloys were tensile tested in vacuum and air as functions of temper-ature from room temperature to 473 K, and the microstructural effect on any moisture-induced embrittlement studied. It has been concluded that:

1. All the alloys and microstructures showed reduced tensile strength (or elongation) in air at room temperature. Tensile strength (or elongation) of the specimens deformed in air tended to recover to the values of the specimens deformed in vacuum, as temperature increased.

2. The measured tensile recovery temperatures indicated that the γ grain microstructure was the most resistant to moisture-induced embrittlement, and the dual phase microstructure was the most susceptible to moisture-induced embrittlement.

3. The moisture-induced embrittlement of the specimens with a fully lamellar microstructure was dependent upon the lamellar spacing, and was reduced with decreasing lamellar spacing.

4. The observed moisture-induced embrittlement of TiAl-based inter-metallic alloys has been discussed in terms of the possible behavior of atomic hydrogen and its subsequent effects on mechanical response in different TiAl-based alloy microstructures.

References

1. Huang, S. C., and J. C. Chesnutt, *Intermetallic Compounds, Volume 2, Practice,* J. H. Westbrook, and R. L. Fleischer, eds., John Wiley and Sons, West Sussex, England, 73, 1995.
2. Kim, Y. -W., *J. Metals,* Vol. 41, 24, 1989.
3. Tsujimoto, T., and K. Hashimoto, *Mater. Res. Soc. Symp. Proc.,* Vol. 133, 391, 1989.
4. Huang, S. C., and E. L. Hall, *Acta Metall. Mater.,* Vol. 39, 1053, 1991.
5. Kim, Y. -W., *Acta Metall. Mater.,* Vol. 40, 1121, 1992.
6. Yamaguchi, M., H. Inui, K. Kishida, M. Matsumoto, and Y. Shirai, *Mater. Res. Soc. Symp. Proc.,* Vol. 364, 3, 1995.
7. Takasugi, T., and S. Hanada, *J. Mater. Research,* Vol. 7, 2739, 1992.
8. Liu, C. T., and Y. –W. Kim, *Scripta Metall.,* Vol. 27, 599, 1992.
9. Nakamura, M., K. Hashimoto, T. Tsujimoto, and T. Suzuki, *J. Mater. Res,* Vol. 8, 68, 1993.
10. Nakamura, M., N. Itoh, K. Hashimoto, T. Tsujimoto, and T. Suzuki, *Metall. Trans. A,* Vol. 25, 321, 1994.
11. Takasugi, T., *Critical Issues in the Development of High Temperature Structural Materials,* N. Stoloff, D. J. Duquette, and A. F. Giamei, eds., TMS, Warrendale, PA, 399, 1993.
12. Liu, C. T., *6th Int. Symp. Intermetallic Compounds—Structure and Mechanical Properties,* O. Izumi, ed., JIM, 703, 1991.
13. Liu, C. T., and E. P. George, *International Symposium on Nickel and Iron Aluminide: Processing, Properties, and Applications,* C. Deevi, V. K. Sikka, P. J. Maziasz, and R. W. Cahn, eds., ASM, 21, 1997.
14. Takasugi, T., T. Tsuyumu, Y. Kaneno, and H. Inoue, *J. of Materials Research,* Vol. 15, 1881, 2000.
15. Kaneno, Y., M. Wada, H. Inoue, and T. Takasugi, *Materials Transaction,* Vol. 42, 418 2001.
16. Stoloff, N., *Hydrogen Effects on Materials Behavior,* N. R. Moody, and A. W. Thompson, eds., TMS, 483, 1990.
17. Chu, W. –Y., and A. W. Thompson, *Meta. Trans. A,* Vol. 23, 1299, 1992.
18. Fitzemeir, L. G., and M. A. Jacinto, *Hydrogen Effects on Materials Behavior,* N. R. Moody, and A. W. Thompson, eds., TMS, 533, 1990.
19. Chan, K., *Metall. Trans. A,* Vol. 24, 1095, 1993.

17

Casting Processes and Simulation Tools

Mark Jolly

CONTENTS

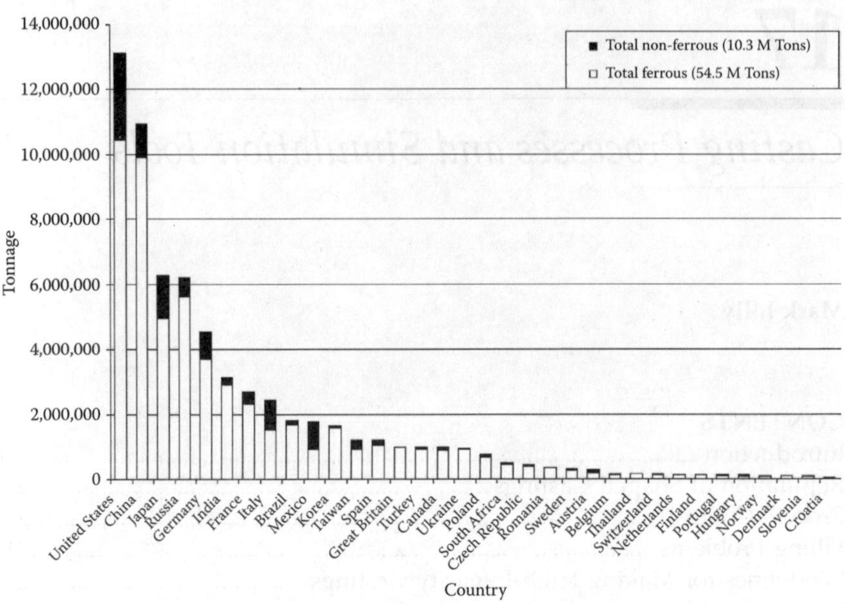

FIGURE 17.1
World casting production for 2000 (data from *Modern Casting*, 2001).

Introduction

The twentieth century has seen the rise and fall of many parts of the foundry industry. Many new alloys have been developed, and with them, new processes. Cast iron continues to play a major role in the foundry sector (Figure 17.1), but its proportion of the market is now generally in decline. In 2000, the total world production of castings was 65 million tons[30] of which it is estimated that over 50% is used in the automotive sector. Although the tonnage of aluminum castings is only 17% of the cast iron tonnage, this equates to some 50% of the volume. The replacement of ferrous-based materials by aluminum has led to a 20% growth in the world tonnage of shape cast aluminum products over the 3 years from 1997 to 2000, from 6 Mt to 8 Mt. In some countries, the growth has been greater. In Sweden, for example, aluminum casting production expanded by 50% from 1995 to 2000. The majority of the expansion has been in high-pressure die castings for the automotive market.

The automotive industry is the biggest user of castings, and the move toward lighter vehicles with better fuel efficiencies has had a serious effect on the mass of cast iron in each car (Figure 17.2). This has had a positive effect on the tonnage of lower density materials, such as aluminum and magnesium alloys. The design of automotive body structures is now being

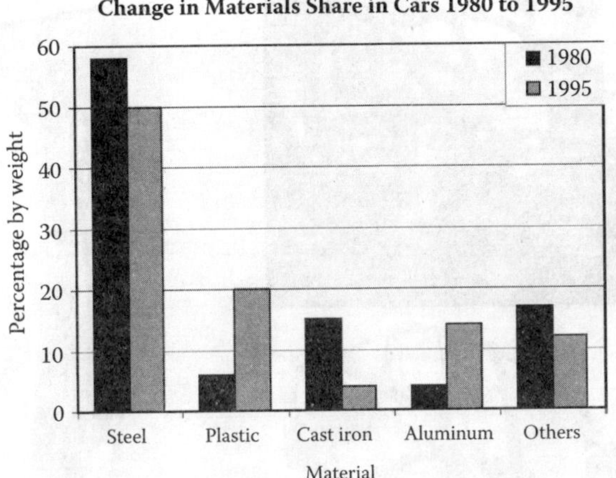

FIGURE 17.2
Chart showing the change in share of materials in cars over the years between 1980 and 1995.

revisited. Every major car producer has now developed an *"aluminum car."* It is actually the body that is aluminum, including the structural components. Some companies have relied solely on aluminum extrusions, but a number have used aluminum cast nodes to join the extruded sections together, and in some cases, castings have been used as items such as B-pillars (Figure 17.3).

Reputation of Shaped Castings

Castings do not have a good reputation within the engineering community. Figure 17.4 shows the scrap rates of component suppliers to a major automotive company indicating that the worst five suppliers are foundries. Often, cast components that are going to be placed under high loads will have a casting factor included in the design that can be as large as 10.[18] In other words, the designer is saying that he can only rely on one-tenth of the published mechanical properties in that cast component. Unfortunately, foundrymen (and it has generally been men) have created this situation by treating the casting process as an art rather than a branch of engineering. Casting is a highly technical engineering process requiring much scientific understanding, and will give sound reliable products if carried out correctly.

On the other hand, designers and engineers are recognizing the advantages that using castings may give: design flexibility, part number reduction by consolidation of fabricated parts into single components, near-net-shape production processes, to name a few. However, these attractive advantages raise other problems associated with design for manufacturing for a process

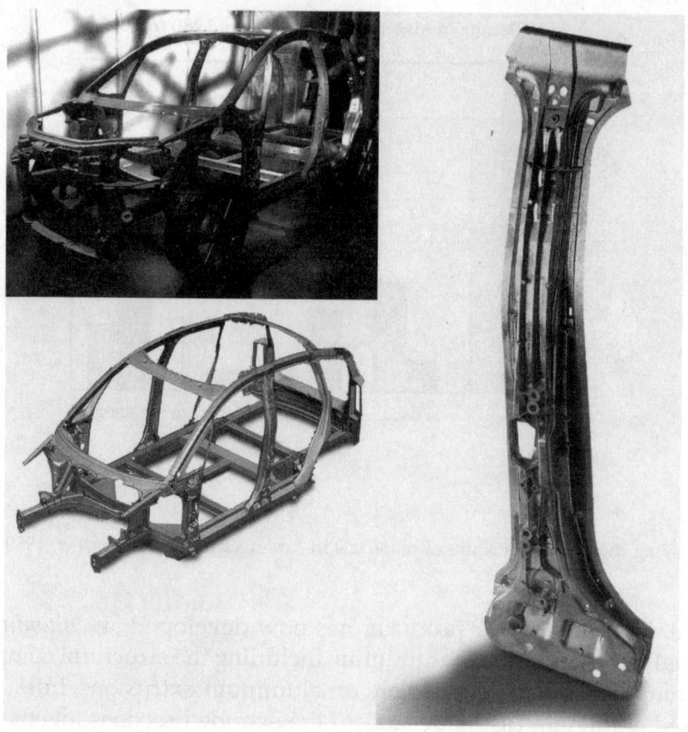

FIGURE 17.3

The Audi A2 space-frame. 22.1% of the A2 is made of castings. The cast parts of the space-frame are marked in red in the schematic. The casting on the right is the B pillar and is a high-pressure die-casting made using a vacural type procedure.

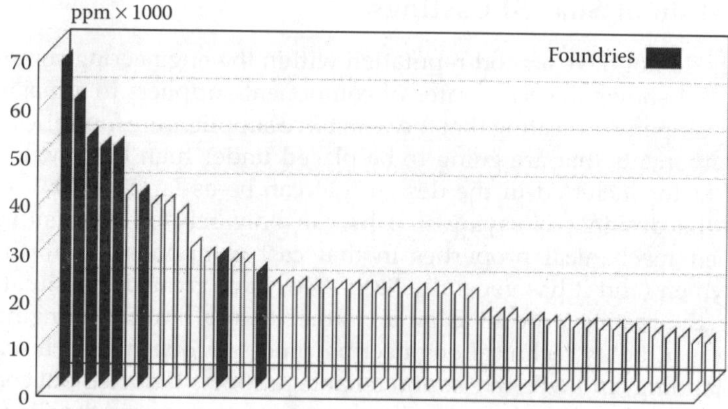

FIGURE 17.4

Suppliers to a major automotive manufacturer in 1995 showing reject rates. Note the five worst suppliers were foundries.

for which a large number of automotive designer engineers appear to have little, or no comprehension. Consequently, foundries are asked to manufacture components that have not been designed with processing in mind, and the poor reputation of castings is then recycled!

Process Control

The level of process control required to produce castings reliably is very high. Composition, temperature of melt and mould, speed of filling, quality of mould/die, and many other factors influence the process, but the main issues that require a good understanding of the science of the process are the quality of liquid metal and the filling, shrinkage, and thermal patterns while the casting is in the liquid state. Castings have mainly been unreliable because they contain porosity. Some of these pores arise from solidification shrinkage, but the majority of them arise from damaged metal. In other words during the processing of the liquid metal, defects have been introduced[5] mainly due to the handling of the liquid metal, that regardless of how one changes the mould or metal temperature will cause the castings to be unreliable under an applied stress.[20] The causes and nature of these defects are discussed below.

Filling Problems

The way in which liquid metal is handled and distributed into a mould cavity is probably the single most important factor affecting the quality and properties of a casting. It is, therefore, strange that the foundry industry and their customers seem to believe that it is the last thing to consider in the design of a casting, and its associated manufacturing process. Work at the University of Birmingham, in the Castings Research Group and the Castings Centre working with industry, and an increasing number of laboratories around the world, has shown that considerable improvements can be made to casting quality by paying attention to the fine detail of the fluid dynamics. When the filling of castings is addressed correctly, the feeding of castings, about which many words have been written, and about which every foundryman has an opinion, becomes a secondary problem.

However, all is not lost! It is heartening to see that a number of equipment manufacturers are designing processes with the quiescent filling of moulds being a prime consideration for aluminum alloys. What now has to be addressed is how to raise the awareness of the rest of the industry to the fact that quiescent, i.e., non-turbulent (not slow) filling is a prerequisite for quality castings. This means all castings, not just aluminum alloys.

Recent work with a number of grey iron foundries has shown that the poor design of filling systems causes most of their scrap. Shrinkage, and therefore feeding, is rarely, if ever, a problem. Slag defects, lustrous carbon films, entrapped bubbles, and sand inclusions give rise to the largest proportion of defects for these automotive suppliers. Such defects only arise from turbulent filling. Traditional iron-gating systems are designed to shoot the metal into a casting by using the gate (the place where the metal passes through into the casting itself) as a choke, which is the place where the metal is traveling fastest. Choke is, therefore, the wrong word! The other traditional view about "the spinner" (a centrifugal trap) cleaning the metal, using the same principal as the spin cycle on a washing machine, is usually completely wrong. The idea is sound, of course, but it is the application of the idea that is incorrect. It has been shown by computer modeling that in the designs of spinners used, the liquid metal takes the shortest path between the inlet and the outlet, i.e., a straight line. At best, the spinner reduces the "yield" (mass of castings made, expressed as a percentage of total metal poured), and at worst, it introduces more slag defects.

Introduction of flat, light, filter-containing running systems into a UK grey iron automotive component supplier has enabled them to start to achieve the very stringent scrap rates imposed by a large, automotive customer. With the original running system, which had been developed over a 15-year period, the scrap rate was a respectable 1.3%. After applying a running system designed using a combination of fluid dynamics, simulation, and practical feedback, castings showed a scrap rate of some 0.3%, of which only 0.15% was due to slag. All the intricacies of controlling molten iron to give slag-free castings are yet to be understood, but the application of sound scientific principles has helped create the step change in defect levels. A drawback, or in hindsight, probably an advantage of the approach, has been that all the other aspects of the process that give rise to problems have also been revealed. For example, control of pouring temperature, placement of filters, glue line procedures, mechanical handling, and mould manufacture have since all come under scrutiny as they are highlighted as causing specific defects that were previously hidden in the overall scrap figures.

Guidelines for Making High-Integrity Castings

Campbell[4] has proposed and Runyoro[37] demonstrated that the overriding feature for good filling of castings is that the metal velocity should never be greater than the critical velocity for surface turbulence. For the majority of engineering casting alloys, critical velocities are in the range 0.4 to 0.6 ms^{-1}. This means that throughout the time when the metal is liquid, it should never be subjected to a velocity greater than this value. This has major ramifications on processes, especially high-pressure die-casting, where the velocity is often as high as 80 ms^{-1}. Even in gravity processes, the metal can achieve velocities of 3–4 ms^{-1} in larger castings, and for Al, the metal only has to fall 12 mm at any point to achieve that velocity. This implies that processes in which the metal enters the mould from the bottom, or in which the mould is rotated, should be inherently better than those where the metal is allowed to fall, or is highly pressurized during the filling. Carrying out Weibull statistical analysis on castings produced using quiescent filling rather than turbulent filling shows that the Weibull modulus can be increased by a factor of between 2 and 5, indicating greater uniformity of properties, i.e., the spread of results is much narrower.[13]

Campbell has published a series of guidelines for the production of high-quality castings and these are often known as the ten commandments.[6]

1. Start with **clean metal** when charging and do not introduce inclusions into the metal.

2. Avoid **meniscus damage** by keeping the metal front velocity below the critical value for surface turbulence.

3. Avoid **liquid front stopping damage** by never allowing the metal front to stop moving during the filling process.

4. Avoid **bubble damage** by preventing the entrainment of air into the casting at any point.

5. Reduce **core blows** by good design and positioning of cores with proper venting.

6. Avoid **shrinkage damage** by ensuring correct feeding, and never try to feed uphill.

7. Avoid **convection damage** by ensuring there are no convection loops created during filling.

8. Stop **unplanned segregation** by understanding and controlling the temperature gradients within the casting.

9. Control **residual stress** by good design, and better heat treatment process—for aluminum avoid quenching in water.

10. Avoid **machining damage** by using a recognized through-process system of datums and location points.

It is impossible to achieve some of these with some manufacturing processes, but the closer the process is to achieving these, then the higher the integrity of castings produced. These guidelines do not consider economic factors.

Most of the work carried out at Birmingham has been on gravity poured systems including sand, shell, investment and metal moulds, or dies, although a limited amount of work has been on low-pressure systems. Recent analyses of high-pressure die-casting failures have revealed that the filling method was the main reason for malfunction of products. It does not matter how high the pressure applied to the liquid is during solidification, once turbulence and turbulence-related defects are created, they do not disappear—they remain within the component, undetectable and unpredictable until a post-mortem is performed, and the design engineers yet again pronounce that castings are unreliable.

Processes

Foundry processes can be split into two main categories: gravity processes, in which the liquid metal is allowed to enter the mould with no externally applied pressure, and pressurized systems, in which pressure is applied to move the metal into the mould/die cavity. When pressurized systems are used, the pressure can also be applied during solidification, in an attempt to improve the soundness of the solid metal.

The physics of casting processes has to deal with extremes. For example, in general for sand casting, except in extremely large castings, the velocity of the liquid ranges from 0.1 to 5 ms^{-1}. However, in high-pressure die-casting, the liquid metal velocity is in the range of 40 to 80 ms^{-1}. Similarly, the externally applied pressure ranges from zero in gravity processes to 250 MPa in squeeze casting.

In the automotive sector, most casting processes are used. Unfortunately, there is still competition between processes for similar parts because no one has arrived at a formula (or one that is accepted) that considers all the variables of the process, including order size, surface finish, part quality, etc. In fact, there is often a problem defining acceptable quality.

Sand casting can be used for the majority of metals. Even highly reactive magnesium is sand cast provided care is taken and the correct materials used.

Sand castings inevitably have a slow cooling rate because of the large insulating mass of sand surrounding the liquid metal as it cools. *Grain sizes* and *dendrite arm spacings* tend to be larger than in the equivalent section sizes in die-castings. Cooling rates in sand castings range from 10^{-6}°Cs^{-1} to 1°Cs^{-1}, and the expected dendrite arm spacing may be from 50 μm to 5 mm.[22] Surface finish for sand castings depends on the fineness of the sand and the quality of the tooling, but is generally poorer than die castings or those produced with externally applied pressure. Minimum section thicknesses achievable in non-pressurized sand castings are about 2–3 mm in aluminum and 3–4 mm in iron.

In *permanent mould casting* (gravity die-casting in the United Kingdom), the component shape is formed from a female cavity, the die of which is usually made of steel or cast iron. Dies are usually sculpted rather than monolithic. Cores can be incorporated that can be manufactured from sand or metal. Inserts of different materials that remain in the final component (for example, threaded bushes for bolts, or stainless steel cooling tubes), can also be incorporated. Pouring of gravity die-castings can be automated in large automotive foundries, but is often manual. Some foundries also tilt during pouring, in an attempt to try to control the metal flow. However, recent work has shown that controlling the flow by tilt pouring requires complex control over the rotation rate, and the accelerations and decelerations involved.[21]

The surface finishes of permanent mould castings is a little better than for sand castings, and is heavily influenced by any coatings that are used and by the position of the surface in the mould cavity. The cooling rates for the process will give dendrite arm spacings in thinner sections of 5 μm to 50 μm. Although some foundries have used the permanent mould process for casting iron components, this is relatively rare; it is more usually used for non-ferrous casting alloys, especially of aluminum. In these alloys, the most common uses are for automotive cylinder heads, pistons, blocks, and castings, where a *heat-treatable casting* is required, often with a specified dendrite arm spacing. Minimum section thicknesses for aluminum alloys in permanent mould casting are 4 to 5 mm. This is determined by the solidification time of the section.

In the *resin shell* or *Croning process*, the mould is made from sand, which is pre-coated with a phenolic resin, blown onto a metal pattern plate (the positive), and cured in an oven. The resulting mould is a thin (up to 15 mm) shell of sand, which usually has a very fine internal surface finish. The moulds can be cored, and chill inserts can be used in order to vary cooling rates. Insulation can also be put on the outside of the shell, if necessary. The split line, of which there can only be one, can be either horizontal or vertical. Pouring can be either automated, semi-automated, or by hand, although there is a predominance of hand pouring in foundries using this process. Filters can be incorporated into the running system. During pouring, and subsequent solidification, the mould may be placed in a container of steel shot (backing), or on a bed of sand, or straight onto a conveyor track.

The surface finish achieved by resin shell casting is generally superior to sand casting, as is the dimensional accuracy, although this depends on the mould–making process. The process tends to be used for higher melting point alloys, such as iron, steel, and copper. Components are usually in the kilograms to tens of kilogram size range, and examples of automotive components are cam-shafts, crank-shafts, tappets, buckets, shims, and slippers. It is often used where chilled white iron is required.

In *lost foam casting* (*evaporative pattern process*), a male copy of the component to be cast is manufactured in polystyrene by blowing into a female die cavity. The polystyrene pattern is then coated with a thin ceramic slurry and dried. The pattern is then placed in a box, and loose sand without a binder is poured around it. The box is usually vibrated to increase the packing density of the sand, and sometimes a vacuum is applied to the box to increase its firmness. Metal is then poured on to the pattern, which decomposes into gaseous materials as it comes into contact with the molten metal.

Manufacture of the patterns can be highly automated, using blowing and steam expansion processes. However, after being removed from the die, the patterns are usually dimensionally unstable and, to obtain stability, have to be conditioned in a temperature-controlled room for a number of days.[43] Pouring can be manual or completely automated. Complex shapes can be produced, without the use of cores, by gluing (often hot melt) smaller bits of polystyrene together. The coating permeability is a significant factor in the process, and therefore, an understanding of the properties of the coating is essential for controlling and understanding it. Recent work at the University of Birmingham has shown that, under certain circumstances, a highly unstable interface can be created between the liquid metal front and the retreating polystyrene as the gaseous products attempt to escape. Other results indicate that in thick sections, the variation of density in the polystyrene can significantly change the way in which the liquid metal flows. Other problems arise from the hot melt glue lines. These materials have different combustion characteristics. The glue bead also gives rise to additional metal in the area of the join that can affect the feeding characteristics of a section.

Despite these inherent problems, the process gives some economic advantages and has been used in a number of plants around the world for long series casting runs. An example is in the GM Saturn engine plant in Tennessee, where the process is used for aluminum heads and blocks, and ductile cast iron differential cases and crank-shafts.

The low-pressure die-casting process consists of a dosing furnace that can be pressurized, and a steel or iron die into which the liquid metal enters from the bottom. It is common practice for low-pressure die-casting foundries to use two levels of pressure; one to fill the mould quiescently, and the second, at a higher pressure, to help feed the liquid metal and reduce shrinkage porosity. The process is usually only used for aluminum or magnesium alloy castings, and is commonly used for automotive wheels. However, some copper foundries have used the process, sometimes combined with graphite dies, for producing components such as taps (faucets) and other domestic water system fittings. Sand cores may be used, and also inserts. As with gravity foundries, most LPDC foundries do not have good temperature controls, and most process cycles are controlled on a time basis. Figure 17.5 shows a photograph of castings produced by low-pressure die-casting.

Low-pressure die-casting is used for medium to long series casting runs, where better mechanical properties are required than are obtainable by conventional high-pressure die-casting. The application of pressure during solidification, albeit only at low levels, is enough in a well-designed component to remove shrinkage problems. However, if the pressure cannot be transmitted hydraulically though the liquid metal to the last solidifying region, shrinkage can still occur, as shown in Figure 17.6. The geometry and alloy type can influence whether the process is capable of making sound components.

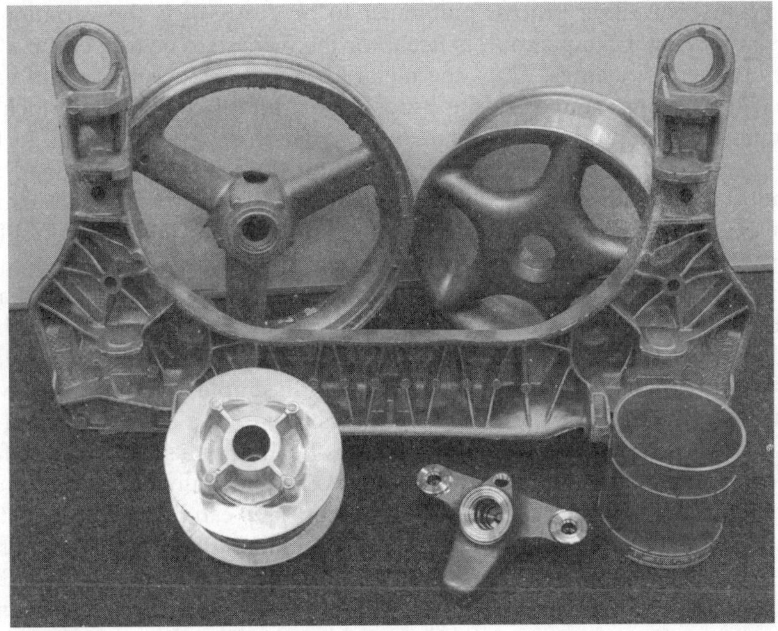

FIGURE 17.5
A selection of low-pressure die-cast components including a motorbike wheel, a car wheel, automotive sub-frame, and wheel hub. (Photo M. Turan)

Heat-treatable components can be produced with reasonable mechanical properties, but there are inherent problems in much of the equipment used to carry out the process, and these lead to scrap levels within foundries that are not often below 10%.

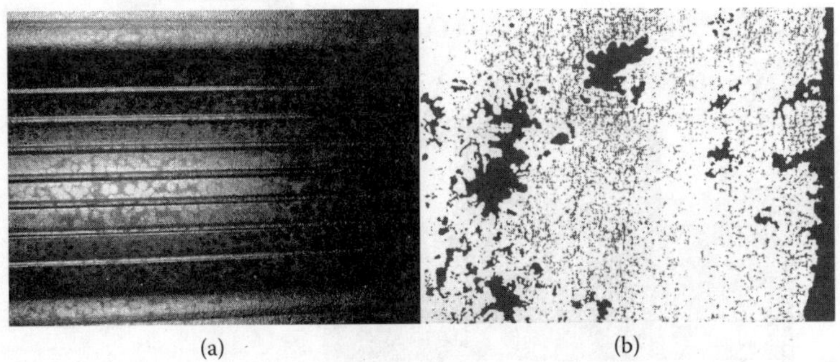

(a)   (b)

FIGURE 17.6
Porosity in a low-pressure die-cast automotive wheel: (a) macro view; (b) micrograph. The porosity was interconnected through the wall thickness.

A major controlling process parameter to be aware of is the variation of pressure during a campaign. It is usual for the furnace to be topped up after a set number of castings. Thus, the metal level within the furnace will vary within the casting campaign, the head height difference being as much as 750 mm. This can produce a pressure head variation of some 170 mbar. As the over-pressure applied is usually in the range from 500 to 1000 mbar, the metallostatic pressure head can have a significant bearing on the total pressure.

Low-pressure sand casting (*Cosworth type processes*) is a combination of LPDC and sand casting. The main difference between this process and gravity sand casting is that, in the Cosworth process for example, the metal is usually pumped electromagnetically into the mould from the bottom. Pressure is then applied until the in-gate is solid. Most other aspects of the casting process are the same as gravity sand casting. This apparently simple difference, however, produces a process with a much higher capability (of the casting meeting its design purpose). At no stage in the process is the liquid metal allowed to fall, and consequently the level of oxide generated is potentially far lower than in a gravity-poured system. An additional process step can be included, that of rotating the whole mould through 180° after filling, although this stage has to be modeled, as it dramatically affects the way solidification occurs. The Cosworth process is shown schematically in Figure 17.7. The process ensures that only clean metal is taken into the mould cavity. The filling profile can be programmed to take account of the geometry of the component, and thus ensure that no liquid front stopping occurs as sections change.

The Cosworth process is really only applicable to aluminum foundries at the moment. It is used by Ford in North America to produce cylinder blocks, and Audi in the United Kingdom to produce cylinder heads. Other low-pressure

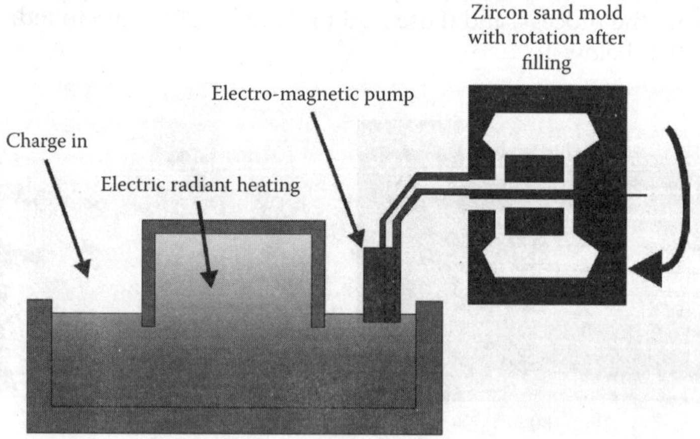

FIGURE 17.7
Schematic of Cosworth process used by Nemak Windsor, (formerly Ford Windsor) Canada, to make cylinder blocks and Audi, Worcester, United Kingdom, and Murray & Roberts, South Africa, to make cylinder heads.

sand-casting processes are used by a number of other foundries to produce automotive components. DISA has made developments to its vertical split line sand-moulded systems (DISAMATICS) to introduce molten metal into the base of the sand moulds. Heinrich Wagner Sinto Maschinenfabrik has also applied low-pressure filling to horizontal spilt line sand moulds.[14] Powertrain Ltd. (United Kingdom), a company that was previously part of BMW, claimed that the box yield obtainable by its LPS process is of the order of 90%–95%. The *Alchemy* process was developed by the BAXI group in the early 1990s to deliver metal into the mould cavity in DISA-type systems.

The *Hitchiner process* was developed at Hitchiner's Technical Center in the mid-1970s, and is a counter-gravity investment casting process. In the counter-gravity low-pressure air melt process (CLA process), the mould is placed in a vacuum chamber with an open snout, or fill pipe, facing down over a furnace of liquid metal. The vacuum chamber is then sealed around the snout and lowered a precise distance into the melt. A vacuum is then generated, which results in the metal being pressurized by the outside atmospheric pressure into the cavity. After a brief holding time during which the components and a portion of the gates solidify, the vacuum is released, and the residual metal in the central snout flows back into the melt. Only a short gating stub remains on the casting, to be removed automatically by machine. The CLA process allows between 60%–94% of the metal to be used to produce a product, compared with 15%–50% in gravity-poured investment processes, where much of the cast weight is in the sprue and gating. The CLA process has traditionally been used only for ferrous-based materials, but it has great potential for aluminum castings. The *CLV process* is a variation of CLA, where the metal is melted under a vacuum and so can be used for more reactive metals. Figure 17.8 shows the CLA and CLV processes.

Unlike gravity-poured parts, which must be cut away from the central sprue, there is no need to leave room for the cut-off blade in the design of a CLA casting cluster. As a result, many more parts can be assembled on a CLA sprue. The increase in pattern population per sprue may be two or three times greater than conventional assemblies. Depending on component size and configuration, this has obvious economic ramifications.

The CLA process provides the ability to cast sections thinner than 0.5 mm. Because the sprue is filled in a non-turbulent fashion from clean metal beneath the surface of the melt, castings with far less slag and non-metallic inclusions are produced. Typically, counter-gravity cast metal contains only 15% of the inclusions of poured metal of the same analysis. This cleaner metal has been shown to reduce tool wear by 100% to 500% in comparative machining tests done under controlled conditions. Because of this and the economic benefits of the CLA process over conventional investment casting, it has started to find its way into foundries producing components for the automotive sector.

Die-casting or *high-pressure die-casting* (HPDC) is the process whereby liquid alloy is forced at high velocity (up to 80 ms^{-1}) and pressure into steel dies.

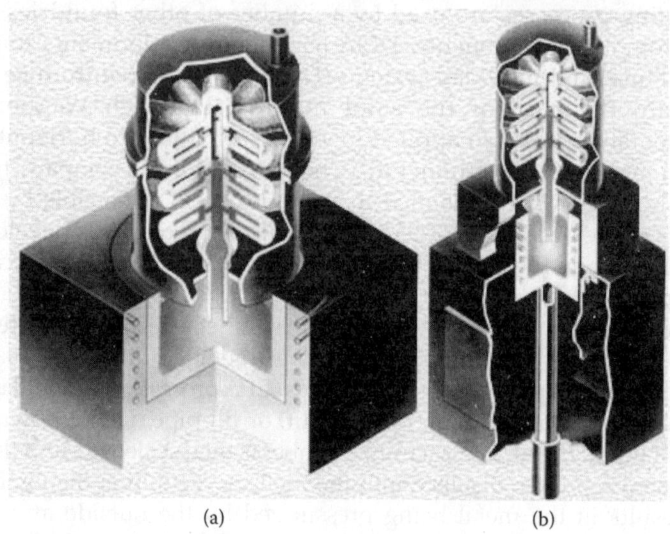

FIGURE 17.8
Hitchiner CLA (a) and CLV (b) processes. In the CLA process the metal is air-melted and in the CLV process the metal is melted under a vacuum.

As with LPDC there is usually at least a two-stage pressure cycle of injection and then intensification. Some of the more modern machines can be programmed to have extremely sophisticated pressure cycles that depend upon the geometry of the casting. Clamping forces on the dies can be from 50 to 5000 tons. Dies are usually monolithic and contain cooling and/or heating channels. Filling times are measured in tens of milliseconds. Cores are not usually present, as most core materials will not withstand the high-impact forces that occur. In conventional HPDC, the metal entry gate freezes very quickly, after which time the casting cavity cannot be influenced by the intensification pressure.

There are two types of HPDC machines, *cold chamber* and *hot chamber*. In general, Zn, Pb, Sn, and Mg alloys castings are made in hot chamber machines, and Al, Cu, and even some ferrous-based alloy castings use cold chamber machines. High-pressure die-casting can produce components that often require little or no fettling, or finish machining. The surface finish and dimensional accuracies achievable are also superior to almost any other casting process. For example, the process is capable of producing castings routinely with wall sections of 0.5 mm in Zn alloys.

The problems associated with high-pressure die-castings are always those of metallurgical soundness and heat treatability. The very nature of the filling process produces a noncoherent liquid metal stream that initially creates a very thin shell of sound material around the surface of the casting. The remainder of the volume of the casting is then filled with a mixture of gas and alloy. Generally, as the gates used are very thin in section, these freeze rapidly before the bulk of the casting is solidified. During the subsequent solidification, which occurs under ambient pressure, gas bubbles can expand and solidification shrinkage is free to occur. Some dies now incorporate additional *squeeze pins* to apply further pressure in areas of the casting prone to shrinkage, and these are activated during the solidification phase. As the castings are often essentially a skin with a core of metal and gas, they cannot be heat treated effectively.

A number of new developments within the industry have been aimed at overcoming the inherent disadvantages of the process. These start with control over the movement of liquid metal in the shot sleeve before it arrives at the gates so that metal in the shot sleeve is not damaged by turbulence. Pressure controls throughout the cycle have also made the filling more con-trolled, but generally the velocities are still very much higher than the critical velocity for surface turbulence. The *Vacural* process, developed by VAW (Ritter and Santarini 1988) attempts to ensure that there is less air in the casting by evacuating the die cavity just prior to injection. This results in less air to entrain, and better filling, as there is no back pressure. The *pore free process* was developed in Japan and used by ASAHI Aluminum.[1] In this process, the mould is filled completely with pure oxygen so that any bubbles formed during the filling will be just oxygen. The theory is that Al_2O_3 is formed, consuming all the available oxygen, and thus all the bubbles disappear.

The perceived potential advantages of high-pressure die-casting drives the industry to seek greater control of the process. Although the process has the potential to be completely automated, and thus, give a high-process capa-bility from an engineering standpoint, there are many metallurgical issues that have to be resolved before the full benefits are achieved.

In the *squeeze-casting* process, liquid metal is introduced into a die cavity, either by direct-pour or via a simple running system. High pressures (up to 250 MPa) are applied to the liquid metal so that solidification takes place under applied pressure. Dies are usually relatively simple and sand cores are not usually used. There is usually no extra metal needed to supply feeding requirements so, theoretically, the yield can approach 100%.

Metal delivery to the die cavity is the weakest part of the process, as the dies are difficult to fill from underneath and then seal against such high pressures. A number of methods are now available by which treated, "undamaged" liquid metal can be introduced into the die. These include the use of *eyedropper ladles* and liquid metal pumps. There are now a number of commercial squeeze-casting machines available on the market.

As well as potentially giving a 100% yield in poured metal, the process enables solidification shrinkage to be eliminated. Cooling rates are high, as there is intimate contact between the casting and die throughout solidifica-tion. Consequently, fine grain sizes and dendrite arm spacings are achieved, even in thick sections. The main drawbacks of the process are the long cycle times, which can be as high as five minutes for castings containing thick sections, and the inability to incorporate cores into the geometry.

Simulation Tools of Casting Processes

The majority of effort in modelling casting has been directed at the filling and solidification stage of the process. In the shape-casting process, the *method* is the term used to describe the extra metal added to the component shape in order to make a casting. The *methods engineer* defines how the metal is to be introduced into the mould cavity (i.e., the *running system*), and the size and positions of the liquid metal reservoirs for feeding (*feeders* or *risers*). In most foundries, the ability to perform this function well comes with experience. Much of the methoding that is performed is highly subject to trial and error.

A number of analytical solutions have been proposed in the past in order to help the cast metals engineer design for feeding, and thus avoid shrinkage. Those most well-known have been proposed by Chvorinov.[7] who defined the concept of "modulus" (the ratio of volume to cooling surface area), by Wlodawer.[50] for directional solidification, and by Niyama et al.[35] For the prediction of centerline micro-shrinkage.

Over the past fifteen years or so, there has been a lot of work to develop computerized simulation programs of the casting process. These simply reproduce the trial and error methods on a computer rather than on a real casting. This approach is potentially both quicker and cheaper, and also enables a better understanding of metal behavior to be obtained. The majority of the software packages developed have been devised by a combination of computer, mathematics, and materials specialists who had little or no knowledge of foundries and foundrymen, and little practical input. As a result, although there are many software packages for the foundryman to use, fundamental misunderstandings still exist as to their usefulness and suitability within the foundry.

Numerical Techniques

Physics and Mathematical Solutions

There are a large number of physical processes that need to be modelled to cover all the shape casting processes. These include heat transfer; including *radiation, convection, and conduction*; mass transfer (mainly fluid dynamics); phase transformations, including solidification and subsequent solid state changes; stress/strain behavior; and microstructural development and segregation of chemical species. The mathematical models used in software codes must also take into consideration the conservation of mass, momentum, and energy.

A wide array of mathematical tools is applied to the physical models in order to solve the physical equations. These come in various guises and combinations of solutions, and include *finite difference* (FDM) and *finite volume* (FVM) methods, *finite element methods* (FEM), *cellular automaton* methods

(CA), and lately, *phase field theory*.[40] Sometimes, there are also combinations of two techniques, such as the *cellular automaton finite element* (CAFE) method proposed by Rappaz and his coworkers.[10] Some simple definitions of the two main techniques used in macro process modeling are given below.

Finite Difference Methods

The *finite difference/volume (FD/FV) method* is the title given to a mathematical technique whereby the answer to a complex problem is obtained by dividing up the complete region (known as a domain) of the problem into small pieces (*control volumes*), and then applying equations to each of these pieces in turn. For each small volume, the calculation assumes that the material properties are the same over the complete volume. Consequently, for high accuracy, the domain of calculation must be split into the highest number of cells possible, or practicable.

Commercial software packages usually apply smoothing algorithms to the results obtained to remove the steps that occur as a result of the small differences between adjacent control volumes. Thus, the results are adjusted in a post-processing step to give smooth contours. The finite difference calculations are carried out at predetermined time steps on an iterative basis. Results can be stored at the end of each time step, or after a predetermined number of time steps. In general, finite difference methods have been used to solve fluid dynamics problems where there is a large amount of material movement.

Finite Element Methods

The finite element (FE) method again divides the domain of interest into many parts, but this time they are called *elements*. The materials properties used in the mathematical calculations are stored at the corners (nodes) of each element, and sometimes, at other places along the edges of the elements for greater accuracy. Solutions of a set of equations are then obtained using these values to give a quantity (for example, temperature) for these specific points (Gauss points) within the element. The positions of these points within the element are determined by the type of integration applied, the initial coordinates of the nodes, and the shape of the element. Unlike the finite difference methods, the values of the variables used in the calculations are not considered constant across the element, but are calculated by using an interpolation method. Time is taken into account iteratively, and step-wise. In general, finite element methods have been used to solve solid-state problems.

Differences between Finite Difference and Finite Element Methods

It has been demonstrated that finite element and finite difference based programs can both produce simulations of filling and solidification processes to the same order of accuracy, and are both capable of dealing with unstructured meshes. However, a recent survey by Jolly et al. (1996) showed that over 50% of the programs used are finite element based, which may indicate that these are more popular, especially in the development stage.

One major factor that has hindered the commercialization of finite element packages has been mesh generation. In finite difference packages, this has always been relatively simple, but automatic meshing for finite element codes was not commercially available until 1995.

Heuristic Techniques

These software packages can cause some difficulties in their use because for commercial reasons, they can produce results without clearly exposing the methodology used. Some of these approaches use a combination of home-grown rules (or *encapsulated knowledge statements*), and some well-known criterion functions, such as the application of Chvorinov's rule and modulus calculations. Others are more complex and less driven by geometry. For example, one package predicts primary shrinkage and porosity in castings by first simulating heat flow through a finite difference type mesh in a full three-dimensional domain, and then by applying experiential criteria to the results. Because this approach requires experimental data for each alloy/process combination, it has to be calibrated for each foundry.

With a number of these packages, the results are extremely good, especially where the software has been well calibrated for the alloys being cast. Even more impressive is the speed with which such software can produce results. Many of the software packages can produce results for 100 million control volumes in a few minutes. More sophisticated software packages may also incorporate a quick solution module that is usually rule-based.

Meshing or Discretization

The action of dividing up the three-dimensional (3-D) solid geometry of a casting that is to be simulated is termed "discretization" or more commonly "meshing." Meshing is one of the most important aspects of the numerical methods approach and it can heavily influence the results obtained during the analysis cycle. Dividing up the 3-D geometry into discrete domains gives rise to numerous problems. For example, representing curved surfaces with cubes inevitably gives rise to steps that hamper any calculations that involve radiation, as the radiating surfaces are all at right angles to each other. Dividing up the object of interest into tetrahedrons, or irregular six-faced cells (hexahedrons) gives a more realistic model for the surface, but gives rise to much more difficult mathematics, especially when dealing with con-servation of mass or volume. However, despite the complex mathematics, in general, stress and flow problems are dealt with better with the less rigid mesh structures, such as those that are hexahedral based. There are still concerns over the problems of so-called "mesh diffusion" when using a tetrahedral mesh, which is the most common form of automatically generated mesh for finite element codes. The usual method for representing curved surfaces is by classing a cell that is more than half full of material as full, and a cell that is less than half full as empty (usually termed marker and cell method, MAC). However, one finite difference-based code, Flow-3D, from Flow Science Inc., Los Alamos, uses advanced mathematical algorithms to represent the free surface of both fluids and curved solids in a rectangular prismatic mesh. This method, termed volume of fluid, or VOF method.[34] models a partially full cell interpolating a surface across the cell in a way that is dependent upon the amount of material in the adjacent cells. This gives an excellent representation of the free surface.

Thermo-Physical Data and Boundary Conditions

One of the most difficult aspects of any simulation exercise is the accurate representation of the boundary conditions. When casting simulation was first used in the foundry industry, most of the codes concentrated on the thermal aspects of the problem. At that stage, there was neither the software nor hardware available to address the problems of filling and liquid metal flow. Boundary conditions were relatively simple, as air temperatures around moulds were assumed to be ambient, and even heat transfer coefficients, although not really known for material combinations, could be estimated for different die materials. In reality, boundary conditions are highly complex, and they may vary considerably through the casting process.

Accurate thermo-physical data input, such as solidus and liquidus temperatures, latent and specific heats, conductivity, viscosity, and surface tension are important for obtaining meaningful results. Some software packages provide the user with a selection of property data for the most common alloys and moulding materials. However, there can sometimes be quite considerable errors in using these data, especially where properties are dependent on some other parameter, such as the time or the solid fraction.

Boundary Conditions and Issues with Casting Processes

At present, many of the codes address filling as well as the thermal aspects, and some also have the capability to model evolving stress. This immediately raises more issues with regard to boundary conditions. If stress models are used, then the heat transfer coefficients change over the time of the simulation as distortion occurs, and gaps are produced between the casting and the mould or die. The initial boundary conditions for filling can also be extremely complex to set up. For example, during a recent investigation at a UK foundry, it was observed that while semi-mechanical pouring of cast iron crankshafts, the metal stream from the lip-poured ladle varied from 20 to 40 mm in diameter, and the distance from the top of the pouring bush ranged from 150 to 250 mm. This led to changes of inlet velocity at the pouring bush from 1.7 ms[-1] to 2.2 ms[-1] during the pouring of one single mould.[48] Similar problems have been pinpointed with bottom-poured ladles, as the head height in the ladle will change between the ladle being full and the ladle being empty. Typically, in a steel foundry, this may mean the exit velocity in the ladle nozzle can vary by as much as 4 ms[-1] from the beginning to the end of pouring.[2] The shape of the metal stream will also vary and depend on whether the metal is lip-poured or bottom-poured.

When simulating any of the die-casting processes, it is usually necessary to impose an initial uniform temperature on the dies, and then apply several (up to 10) solidification cycles in order to produce a temperature profile for the steady state condition within the die before completing a full analysis of filling and solidification. Modelling of coatings (or paints), which may be no more than 0.1 mm thick, is not really possible at present, and so, in order to influence the amount of heat transfer across any coating, a heat transfer coefficient has to obtained for the coatings used, and applied to the surfaces that are coated. Currently, there is no full, validated database of interface heat transfer coefficients for different materials combinations.

The major issues with regard to modelling of the investment casting process start from the geometrical considerations. Often, the aspect ratio of thinnest section to overall size of the casting is large. This, therefore, gives problems during the meshing stage. Other issues relate to the proprietary nature of the shell materials. Each foundry may use its own combinations of materials, and therefore there is no database of shell material properties that can be usefully created for the foundry community. A method of predicting some of the properties required for modeling is being developed by Jones and coworkers.[23] An insulating blanket is sometimes used in an ad hoc fashion to reduce the rate of heat loss in certain areas, but chills are not usually used. Radiation is a major consideration in this process, and so the simulation approach must consider this aspect of heat flow as well as conduction. In resin shell casting, a backed mould must be modeled differently from an unbacked mould, where radiation plays a considerable role in the heat transfer processes.

In lost foam casting where pattern parts are glued together, there are problems in modelling the glue material that usually has very different properties form the bulk pattern properties. Other issues arise in trying to model the decomposition of the polystyrene. However, work in the United States is gradually addressing this problem.[31] Other results indicate that, in thick sections, the variation of density in the polystyrene can significantly change the way in which the liquid metal flows, and also influence the local thermal conditions.[24]

Modelling tilt casting adds to the problems associated with either sand or die-casting. Knowing how the foundry carries out the tilting, in terms of the equipment stiffness and the angular tilt rate, is extremely important.

When modelling high-pressure die-casting, the engineer has to be aware that, unlike in the other casting processes, the metal stream probably does not stay coherent, and is more likely to be like a spray of particles. The air in the die is hugely influential on the filling pattern. Venting of the die is, therefore, essential, and should be modeled. Although sand cores are not used in this process, some of the dies used are extremely complex, and the section thicknesses can be as small as 0.1 mm. Sometimes, additional post-filling pressure is applied by means of "squeeze pins" to specific locations in the casting, and this is not easy to model.

Thermo-Physical Data

Exothermic materials. It is common practice within steel and some iron foundries to use exothermic sleeve materials for feeders/risers or "hot toppings." These are manufactured from materials that produce a thermit type reaction when in contact with liquid metal. These are proprietary materials that put heat back into the riser, thus enabling the liquid metal to stay liquid for longer. Only one serious attempt at measuring these properties has been made, and the results to date are proprietary.[28] A number of software packages do have the capability of dealing with these materials, provided that the thermo-physical data are available, although there are a number of different ways in which the materials are treated.

Nonstandard alloys. Exotic or proprietary alloys can often cause a problem for the foundryman as there is no universal database of thermo-physical data for general access. Data for standard alloys now are relatively easy to find. Data for alloys with slight deviations from standard, or exotic alloys with low annual production tonnage and potentially with more casting problems, do not exist in the public domain, and are extremely expensive to produce. There are, however, a number of software packages that can be used to produce calculated thermo-physical data from chemical compositions, such as Thermotec and MTData.

Mould material properties. Thermo-physical data for moulding material properties, especially for sand moulds or investment shell moulds, are not well documented. Problems arise from the very fact that sand properties will vary from foundry to foundry. Investment shell technology is often unique to a foundry, and regarded as a proprietary art. It is difficult to envisage how this issue can be resolved completely in modelling. Some headway is being made by use of some simple rule-based techniques in order for foundries to be able to estimate certain thermo-physical data.[23] Such tools could be applicable to other situations in the foundry, for example in mould coatings. Work on the modelling of sand core blowing will also contribute to a better understanding of the effect of cores during the casting process.[38, 49]

Interface heat transfer coefficients. The interface heat transfer coefficient is probably one of the most "fudged" parts of casting modeling. As the mould material properties are not accurately known, it is not possible to make accurate measurements of interface heat transfer coefficients although, some attempts have been made.[25] Often, the mould and the interface are treated as the same, even though there are certain mechanisms occurring that must change the value of heat transfer coefficient during the casting process. As previously discussed, the addition of die coats and mould washes will change the heat transfer coefficient, and this should be taken into account in a more systematic way than is currently the practice. Often, interface heat transfer coefficients are used that enable the user of the software to arrive at the correct defect prediction, and studies are continuing to be able to predict values of heat transfer coefficients.[16]

Viewing Results

Computer simulation software is of no use unless the mass of information that is generated can be easily interpreted by those who are making castings, in order for them to change the process in a structured fashion. With the development of faster computers, modellers are expected to present results in easily digestible images. However, although much of what is seen in computer simulation is still qualitative, work is ongoing to develop quantitative methods of analysis of casting processes using modelling as a tool.[15]

Filling

Relating modelling of filling to casting quality is probably the least well-developed aspect of casting simulation. Filling results can be represented as velocity vectors, velocity contour maps, and temperature distributions (isothermals) in both 2- and 3-D images. Two-dimensional images are cross-sections through the liquid, and are often easier to understand than a fully transparent 3-D model, even though a large number of sequential images is required to convey all the information about a filling system. Images of the free surface behavior can also be useful, although surface turbulence using these images has to be interpreted carefully. It is also possible to highlight all areas of the liquid metal that are above the critical velocity during any time step. Particle tracing methods are also used to give some indication of the progression of fluid, and the development of phenomena, such as eddies or strong flow fields.

A more recent development is the association of a scalar value with the free surface that has time dependence, and is associated with the oxide-generation mechanisms of free surface turbulence. This enables running systems to be evaluated a little more quantitatively. Figure 17.9 shows some methods of representing the filling of castings.

For high-velocity flows, such as those that occur in high-pressure die casting, there is some debate as to whether current finite difference (FD) and

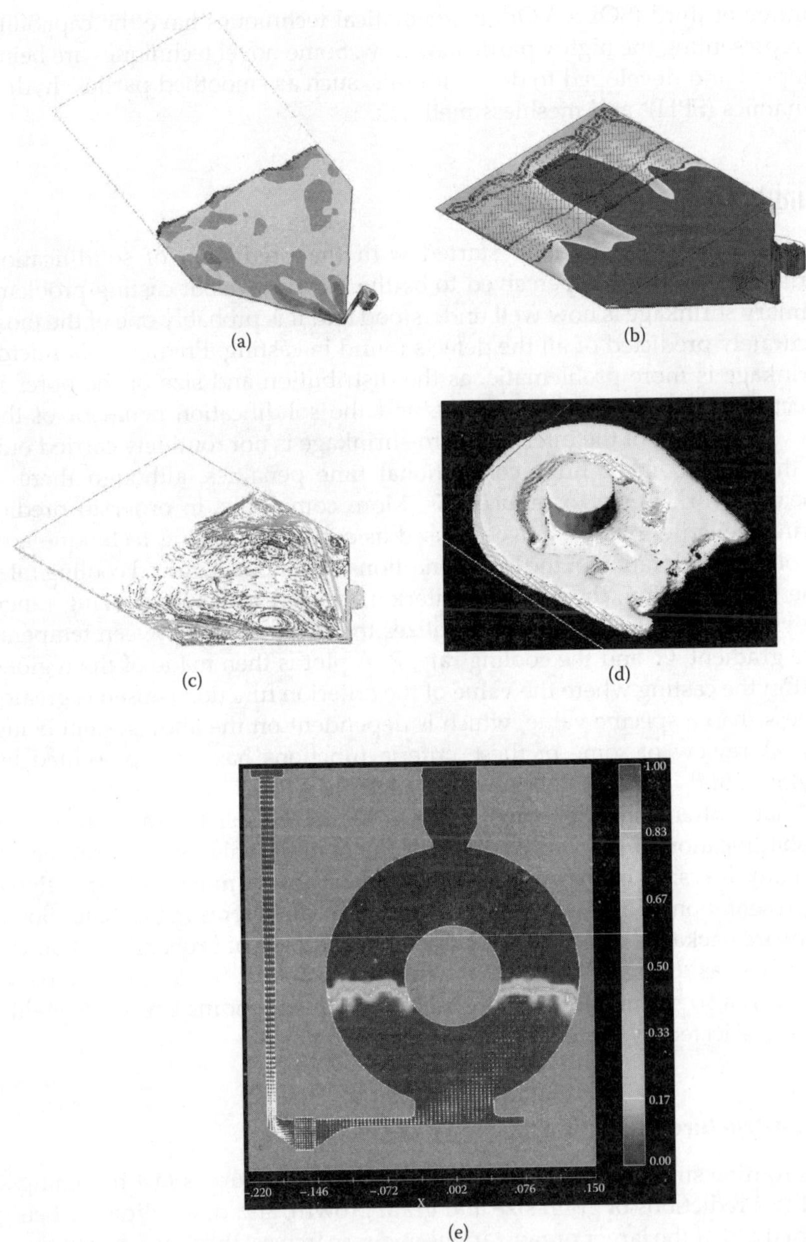

FIGURE 17.9
(a) velocity contours (b) liquid mushy and solid regions during filling of a plate, (c) particle tracking enabling the interpretation of where inclusions might be washed to during the casting process as well as indications of bulk filling history, the development of dead zones and eddies, (d) rendered surface of liquid metal, and (e) is a 2-D cross-section through the liquid metal showing fraction of material and velocity vectors.

volume of third (SOLA-VOF) mathematical techniques have the capability of representing the highly particulate flow. Some novel techniques are being adopted and developed to deal with this, such as smoothed particle hydrodynamics (SPH)[8] and meshless methods.[26]

Solidification

Casting simulation initially started with the prediction of solidification shrinkage, as this was perceived to be the most prevalent casting problem. Primary shrinkage is now well understood and it is probably one of the most accurately predicted of all the defects found in casting. Prediction of micro-shrinkage is more problematic, as the distribution and size of the pores is affected heavily by both the filling and the solidification behavior of the alloy. Modelling of the internal micro-shrinkage is not routinely carried out, as that would incur huge computional time penalties, although there is research work to try to reduce this. More commonly, in order to predict shrinkage, criterion functions are used as a post-processing technique and involve consideration of the local conditions of temperature and cooling rate. One such function, the Niyama criterion, is used for short freezing range alloys, and especially steels, and utilizes the relationship between temperature gradient, G, and the cooling rate, R. A plot is then made of the regions within the casting where the value of the criterion function chosen is greater or less than a specific value, which is dependent on the alloy system being cast. A review of some of these criteria functions has been presented by Taylor et al.[41]

What is often valuable to the foundryman is to be able to predict the order of solidification of the various parts of the casting. This can be represented in many ways: as isothermals, isochronals, or liquid/mush/solid fractions. Representation of porosity has always led to an interesting debate. Some software packages represent the porosity as contours of probability of porosity, others as discrete "holes" in the casting geometry. Some software packages can also show different levels of porosity depending on what quality level is selected by the operator.

Microstructure Prediction

The routine simulation and prediction of microstructure is fast becoming a reality. Predictions of grain size and grain growth, and orientation are being carried out at the larger organizations and research establishments, but these have yet to make their way onto the shop floor of the smaller foundries. Modelling of grain size, and the production of pole figures of texture in aerofoil castings is now possible, as shown in Figure 17.10. Individual dendrite growth can also be modelled well using either cellular automaton (CA)[17] or phase field techniques.[47, 46] Figure 17.11 illustrates what can be seen

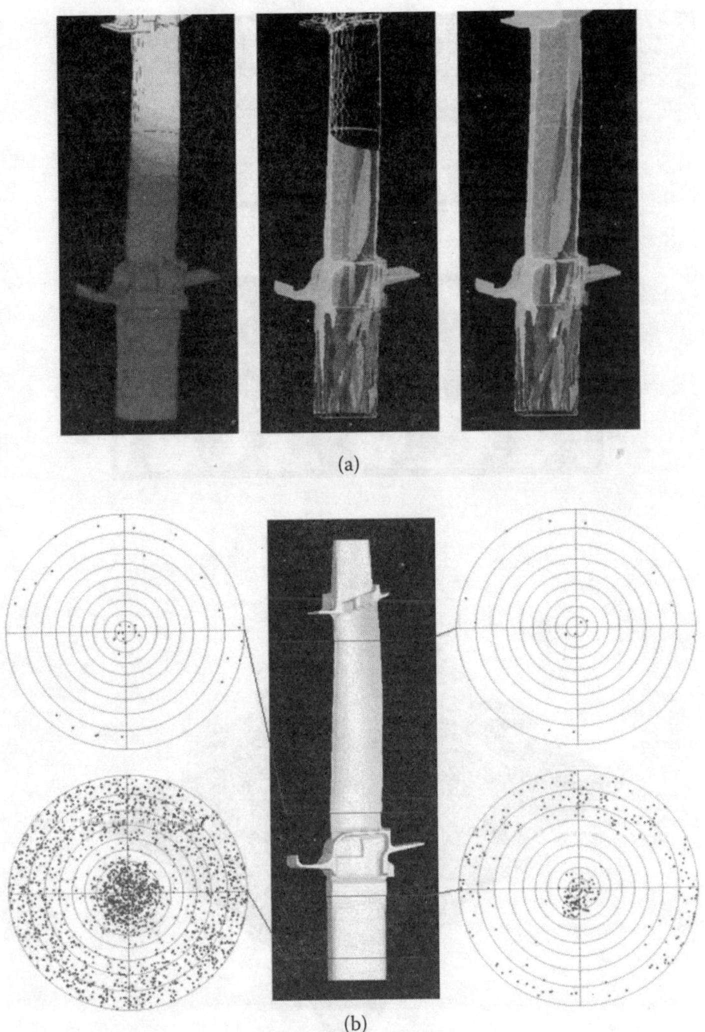

FIGURE 17.10
Prediction of grain structure (a) and orientation with pole figure (b) in a directionally cast turbine blade.

from these two types of analysis. However, there is some way to go yet before this can be carried out on a full-size casting, as the CPU required for discretization at that level is enormous. Voller[44] has predicted that with current rates of improvements in hardware and software, it would still take until 2040 to be able to model dendrite tips in a casting space envelope of $10 \times 10 \times 10$ cm.

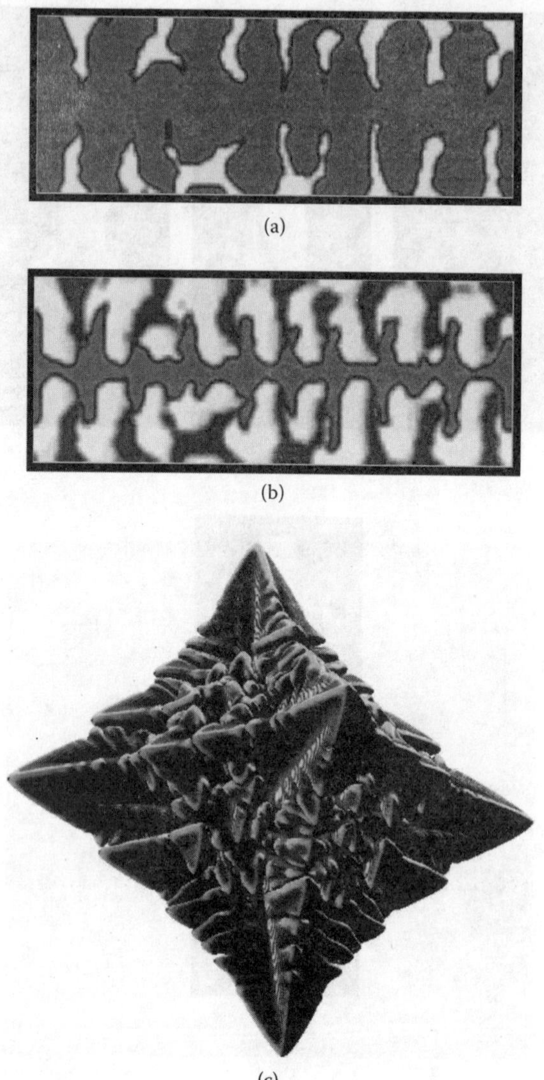

FIGURE 17.11

(a) and (b) development of 2-D dendritic structure and microsegregation predicted using the CA technique (Jarvis et al. 2000), (c) 3-D dendritic growth modeled using the phase field technique.[17]

Mechanical Property Prediction

It is possible to infer mechanical properties from the thermal history of the casting. Hardness and strength have both been predicted using these methods, and images mapping contours can be produced.

Stress, Distortion, and Shape Prediction

Prediction of distortion or shape is extremely important, as this can influence the thermal patterns in the casting and, hence, the final microstructure or porosity. Modelling the gap formation between the casting and mould has been achieved by a number of software packages. Macro-shape change and casting stresses can also be predicted rather well. The propensity for hot tearing can be assessed, although this is only currently possible by interpretation of the data, since a good model for the nucleation of tears does not yet exist.

Other Processes in the Manufacture of Castings

A number of automotive and aerospace companies are adopting a strategy of achieving "through-process-modelling." This means that every part of all processing routes is modelled to enable the variations in process to be understood more effectively and to achieve optimum manufacturing conditions more quickly. For casting, this entails modelling a number of the processes not regularly simulated. Some examples of such processes are: sand mould manufacture.[3, 27] Core blowing investigated by Williams and Snider using a variation of the Marker and Cell technique, and superimposing Lagrangian and Eulerian meshes; and lost foam casting. For investment casting, the other processes to consider are core injection, pattern manufacture, shell manufacture, and pattern removal, and these have all been investigated at the University of Birmingham.[23]

Future

There has been much discussion about potential changes in the automotive sector, with the electric car, the hybrid car, the fuel cell, hydrogen engines, and others besides, all being investigated as alternatives to the internal combustion engine. Any mass-market acceptance of these technologies will have a major impact on the foundry sector, as there are currently a large number of cast components in the automotive powertrain. However, there are growing numbers of castings in automotive structures, steering, and suspension sections. Cast iron will have difficulty in keeping its place within the automotive sector as a major material as cylinder blocks, camshafts, and crankshafts. However, aluminum has made inroads into all automotive areas, and the future lies in producing cleaner metal for cleaner and more reliable castings. The trend for increased investment in process control instruments, melt transfer and handling equipment, is certain to continue.

Summary

This chapter has attempted to show that, for the automotive sector, foundries continue to be a major source of component supply. Changes within the foundry sector are rapidly turning what was an art during most of the twentieth century into a high technology engineering process in the twenty-first century. Research in universities concentrating on understanding the problems associated with castings have led to process developments and introduction of new process controls. It is no longer acceptable to make more castings simply to supply the customer with enough castings of acceptable quality. Understanding the effect of metal flow on internal defects has led to the development of processes such as Cosworth and LPS. High-pressure die-casting is being used as a process of choice for surface finish, and near net shape products, but is also being developed for structural and heat-treatable components for the future. Low-pressure die-casting remains a major player for wheels. Castings have moved into new areas, such as doors, B-pillars, and suspension parts by the combination of component, material, and process design with good process control. Variations of the investment casting process show as-yet undeveloped potential for the automotive sector. Process modelling is now being used by a majority of automotive foundries that combined with 3-D CAD, will give the ability to achieve the automotive sector's aim of through-process modeling.

References

1. Asahi Aluminum, www.asahial.co.jp.
2. Ashton, M. C., and R. Wake, "Evaluation and control of bottom pour ladle practice," *Steel Casting Research and Trade Association Journal of Research*, Vol. 35, 16, December 1976.
3. Bakhtiyarov, S. I., and R. A. Overfelt, "Experimental and Numerical Study of Bonded Sand/Air Two Phase Flow in PUA process," *Trans. AFS*, Vol. 110, 159–180, 2002.
4. Campbell, J., 1991, *"Castings,"* Butterworth Heinemann, Oxford.
5. Campbell, J., "Invisible macrodefects in castings," *Journal de Physique IV*, Coloque C7, supplément au *Journal de Physique III*, 3, November 1993.
6. Campbell, J., "10 Rules for good casting," *Modern Castings*, 1997, 97(4), 36–39.
7. Chvorinov, N., *Giesserei*, Vol. 27, 201–208, 1940.

8. Cleary, P., and J. Ha, "Three-Dimensional Smoothed Particle Hydrodynamic Simulation of Light Metal Components," to be published in *Journal of Light Metals*.

9. Flemmings, M. C., D. Apelian, D. Bertram, W. Hayden, P. H. Mikkola, and T. S. Piwonka, "Advanced Casting Technologies in Japan and Europe," *WTEC Panel Report*, AFS and ITRI, Loyola College, MA, March 1997.

10. Gandin, Ch.-A., T. Jalanti, and M. Rappaz, "Modeling of Dendrite Grain Structures," *Modeling of Casting, Welding and Advanced Solidification Processes—VIII*, B.G. Thomas and Ch. Beckermann, eds., TMS, Warrendale, PA, 363–374, 1998.

11. Gebelin, J.-C., A. Cendrowicz, and M. R. Jolly, "Wax Injection in the Investment Casting Industry," *Proceedings of Computational Modeling of Materials, Minerals, and Metals Processing*, Seattle, WA, ISBN 0-87339-513-1, February 2002.

12. Gebelin, J.-C., S. Jones, and M. R. Jolly, "Modeling of the De-waxing of Investment Cast Shells," *Proceedings of Computational Modeling of Materials, Minerals, and Metals Processing*, Seattle, WA, ISBN 0-87339-513-1, February 2002.

13. Green, N. R., and J. Campbell, "Statistical distributions of fracture strengths of cast Al-7Si-Mg alloy," *Mat. Sci. and Engineering*, A173, 261–266, 1993.

14. "Greensand moulding machine development continues apace," *Foundryman*, 92(9), 261–266, September 1999.

15. Griffiths, W. D., and M. Cox, "Quality indices for modeling of mould filling," *FOCAST 2nd Mini Conference*, University of Birmingham, EPSRC Grant GR/M60101, Paper 10, 2001.

16. Hallam, C., *Interfacial heat transfer between solidifying aluminum alloys and coated die steels*, PhD Thesis, UMIST, UK, 2001.

17. Jarvis, D. J., S. G. R. Brown, and J. A. Spittle, "A 2D cellular automaton-finite difference (CAFD) model of the solidification of Al-Cu-Si alloys," *Proc. Light Metals 2000*, 129th TMS Annual Meeting, 603–608, March, 2000.

18. Jolly, M. R., "Removing the Casting Factor," *Foundry Trade Journal*, 171(3532), 273–274, July 1997.

19. Jolly, M. R., S. Wen, and J. Campbell, "An Overview of Numerical Modeling of Casting Processes," *Modeling and Simulation, Metallurgical Engineering and Materials Science*, Beijing, China, June 11–13, 1996, Proc. pub Metallurgical Industry Press, ISBN No 7-50241908-X/TG, 540.

20. Jolly, M. R., and J. Campbell, "Beating the Cancer Within," *Foundry Trade Journal*, 175(3557), 5, August 1999.

21. Jolly, M. R., M. Cox, J.-C. Gebelin, S. Jones, and A. Cendrowicz, "Modeling the Investment Casting Process: Some preliminary results from the UK Research Programme Fundamentals of Investment Casting (FOCAST)," *Trans. AFS*, Vol. 104, ISBN 0-87433-237-0, 2000.

22. Jones, H., *Proc. of Conference on Rapid Solidification Processing: Principles & Technologies*, Reston, VA, USA, R. Mehrabian, B. H. Kear, and M. Cohen, eds., Claitors Pub Div., Baton Rouge, LA, 28, 1978.

23. Jones, S., M. R. Jolly, and K. Lewis, "Development of Techniques for Predicting Ceramic Shell Properties for Investment Casting," *Br. Ceram. Trans.*, 101(3), 106–113, June 2002.

24. Kahn, S., C. Ravindran, and D. Naylor, "Effect of Casting section thickness and coating thickness on the interfacial heat transfer coefficient in lost foam casting," *Trans. AFS*, Vol. 105, paper 01–070, 2001.

25. Kayikci, R., and W. D. Griffiths, "The influence of surface roughness on interfacial heat transfer during casting solidification," *Castcon '98, Proc. of the Annual Conference of the Institute of British Foundrymen*, 18–19 June 1998.

26. Lewis, R. W., and J. Bonet, *Development of particle based meshless methods for simulation of mould filling in pressure die casting*, EPSRC Research Grant GR/M84312.

27. Makino, H., Y. Maeda, and H. Nomura, "Computer Simulation of Various Methods for Greensand Filling," *Trans. AFS*, Vol. 110, 137–146, 2002.

28. Medea, A., "Accurate exothermic/insulating steel feeding system performance predictions using computer simulation," *Trans. AFS*, paper 98–002, 1998.

29. Mi, J., R. A. Harding, and J. Campbell, "The Tilt Casting Process," *Int. J. Cast Metals Res.*, 14, 2002.

30. Modern Casting Staff report, 35th Census of World Casting Production—2000, *Modern Casting*, 38–39, December 2001.

31. Molibog, T. V., and H. Littleton, "Experimental Simulation of Pattern Degradation in Lost Foam," *Trans. AFS*, Vol. 105, 01–020, 2001.

32. Molibog, T. V., "A mathematical model of molten metal/foam polymer interaction in lost foam casting process," to be published in *Applied Mathematical Modeling*.

33. MTData Division of Materials Metrology, National Physical Laboratory, Teddington, Middlesex, UK.

34. Nichols, B. D., C. W. Hirt, and R. S. Hotchkiss, "SOLA-VOF: a solution for transient fluid flow with multiple free boundaries," *Technical report LA-8355*, Los Alamos Scientific Laboratory, 1980.

35. Niyama, E., T. Uchida, M. Morikawak, and S. Saito, *International Foundry Congress 49*, Chicago, paper 10, 1982.

36. www.ritter-aluminium.de/english/vacural_1.htm.

37. Runyoro, J. J., 1992, *Design of gating systems*, Ph.D. thesis, University of Birmingham, UK.

38. Salmonds, M., Private Communication, February 2001.

39. Santarini, M., "Aluminum castings in the automobile: balance sheet and perspectives," *Hommes et Fonderie*, 286, 12–22, Aug–Sept 1998.

40. Steinbach, I., and G. J. Schmitz, "Direct numerical simulation of solidification structure using the Phase Field method," *Modeling of Casting, Welding and Advanced Solidification Processes—MCWASP VIII*, B.G. Thomas and Ch. Beckermann, eds., TMS, Warrendale, PA, 521–532, 1998.

41. Taylor, R. P., J. Shenefelt, J. T. Berry, and R. Luck, "Comparison of, and Criticism of, Casting Criteria Functions," *Trans. AFS*, paper 02–042, Vol. 110, 315–330, 2002.

42. Thermotech Ltd., Surrey Technology Centre, The Surrey Research Park, Guildford, UK.

43. Vatankhah B., and H. E. Littleton, "Dimensional control parameters in lost foam casting," *Trans. AFS*, 01–106, 105, 2001.

44. Voller, V. R., and F. Porte-Agel, "Moore's Law and Numerical Modeling," *Journal of Computational Physics*, JCPH2001-0324, 2001.

45. Voller, V. R., "Micro-Macro Modeling of Solidification Processes and Phenomena," *Computational Modeling of Materials, Minerals, and Metals Processing*, M. Cross, J.W. Evans, and C. Bailey, eds., TMS, Warrendale, PA, ISBN: 0-87339-513-1, 41–61, 2002.

46. Warren, J. A., and W. George, Center for Theoretical and Computational Materials Science, National Institute of Standards and Technology, http://www.ctcms.nist.gov/~jwarren/, 2002.

47. Warren, J. A., and W. J. Boettinger, "Prediction of Dendritic Growth and Microsegregation Patterns in a Binary Alloy using the Phase-Field Method," *Acta Metall. Mater.*, Vol. 43, 689–703, 1995.
48. Wen, S., N. Hastings, and M. R. Jolly, *Modeling of Filling in Crank Shaft Castings*, confidential report The Castings Centre, IRC in Materials, The University of Birmingham, Birmingham, UK, 1998.
49. Williams, K., D. Snider, M. Walker, and S. Palczewski, "Process Modeling: Sand Core Blowing," *Trans. AFS*, Vol. 110, 237–256, 2002.
50. Wlodawer, R., *Directional Solidification of Steel Castings*, Translation, L. D. Hewit, and R. V. Riley, Pergamon Press, 1966.

18

Damage Tolerance in Composite Structures

Ivana K. Partridge

CONTENTS

Introduction

The intrinsic brittleness of highly cross-linked thermosetting polymers is translated into brittleness of fiber-reinforced composites made from them. Numerous approaches to toughening of thermosetting composites have been researched, many passing into industrial practice. This chapter describes some of the less well-known aspects of toughening of polyester/ chopped glass fiber composites. It also considers the issue of delamination in laminated composites made from carbon fiber/epoxy prepregs. Selectively placed interleaves of the matrix resin between the prepreg plies are shown to be capable of increasing the resistance to crack initiation between the plies, with a strong dependence on the interleaf location and thickness. The technique of Z-Fiber® pinning is outlined, together with its ability to reduce, or even stop, delamination cracks from propagating in a laminated structure.

Background

The studies described here have focused on some of the available approaches to the toughening of fiber reinforced composites having thermosetting polymers as the matrix. This focus is justified on the basis that commercially available thermoplastic polymer composites, such as continuous glass fiber reinforced polyolefins (GMT compounds), or even high-temperature thermoplastics (e.g., polyetherimide used as a composite matrix), already exhibit a relatively high level of damage tolerance. The use of thermosetting composites in the automotive industry covers a wide range, from the classical and inexpensive polyester moulding compounds, to combinations of the most expensive forms of continuous carbon fibers with highly toughened epoxy resins. This chapter presents some very different approaches to increasing the damage resistance of two classical forms of thermoset matrix composite. The first are sheet-moulding-compounds (SMC) involving polyester/chopped glass fiber mat combination, while the second are high-specific stiffness laminates made from epoxy-impregnated continuous carbon fiber tapes.

The use of SMC is specified in the commercial vehicle market because, in addition to its ability to create high style, it lowers vehicle weight, reduces tooling costs, and resists corrosion and denting. As an example, the Cadillac XLR hood assembly includes a complex-shaped, non-appearance SMC inner panel that is bonded to an SMC outer panel with a Class A finish. The hood surface has a four-sided, tapered design and includes steel hinges, brackets, and the hood latch.

The relatively expensive toughened epoxy resins, intended for use with high-strength carbon fibers, were developed specifically for demanding Formula 1 applications. Irrespective of the end of the spectrum of applications, the currently available composite systems still suffer from the classical problem of a tendency to microcracking in the resin, and delamination in the composite structures, falling short of new demands on increased performance across the automotive sector.

Increasing Dent Resistance

This section concerns the resistance to damage of composite parts compression molded from SMC. The matrix resins are almost invariably polyester resins, with or without low profile modifiers and a high filler loading. Polyester resins are highly cross-linked, exhibiting glass-to-rubber transition temperatures T_g in excess of 100°C. This high cross-link density is accompanied by a low fracture toughness G_{IC} of the cured resin, typically below 150 J/m². However, it is the high cure shrinkage exhibited by polyester resins (up to 10% by volume), that is perceived to be the biggest problem. The problem

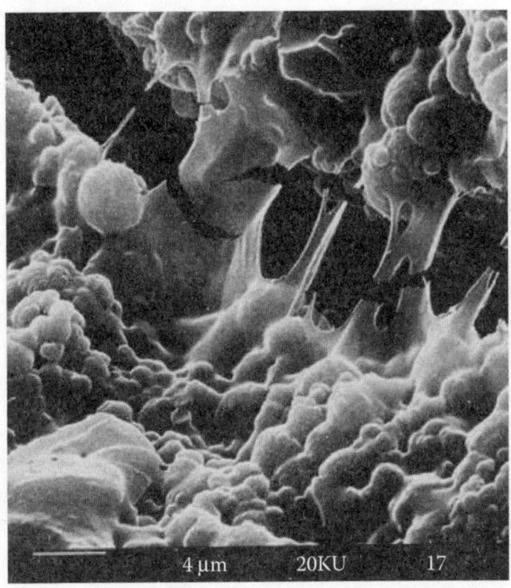

FIGURE 18.1A

Microfissuring in a PVAc modified cured polyester resin. The spherical entities are composed of highly cross-linked polyester resin, with the thermoplastic being the covering, semi-continuous, phase. (The phase identification can be achieved by etching an equivalent fracture sample in a suitable solvent such as acetone, or methyl ethyl ketone.)

of brittleness is reduced, to an extent, when the resin is used as the matrix in a glass mat reinforced composite, primarily by the chopped and randomly oriented fibers acting as effective crack stoppers. The issue of resin shrinkage is then tackled by the addition to the resin of the so-called low-profile modifiers.[1] Polyester resins modified by the addition of a low T_g thermoplastic, such as the traditional low-profile additive polyvinyl acetate (PVAc), achieve their resistance to volumetric shrinkage by extensive microfissuring. This occurs during the cure, at weak interfaces, either at the resin/glass or resin/filler boundaries, or at the interface between the thermoplastic modifier and the polyester,[2,3] as shown in Figure 18.1a for PVAc modified cured polyester resin and in Figure 18.1b for a cured polyester resin filled with hollow glass balloons (Fillite). This is expected to be at the expense of the intrinsic mechanical strength of the cured system. A different type of modifier, namely a reactive liquid rubber, e.g., carboxyl terminated butadiene rubber, (CTBN) has long been used to increase the toughness of epoxy resins in applications such as adhesives.[4] The possibility of using reactive rubber modifiers to act as a shrinkage moderating additive, as well as a toughening agent, in polyesters thus appears worthy of consideration.

The results shown in Figure 18.2a indicate that the shrinkage reduction in a polyester resin, brought about by the inclusion of a CTBN rubber, is comparable to that achieved by addition of a commercial low profile additive (LPA). At the same time, the glass transition temperature of the cured blend

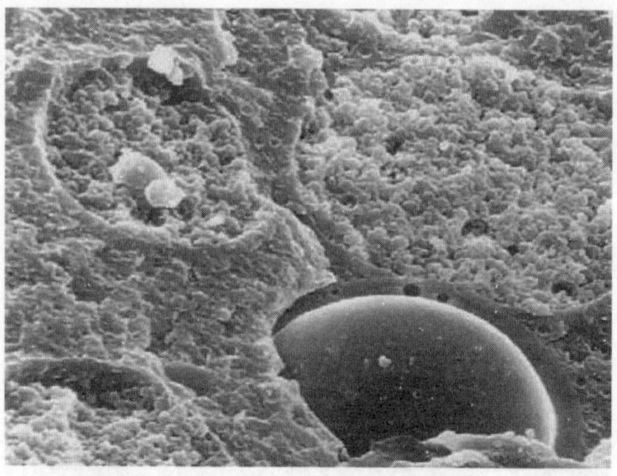

FIGURE 18.1B

Scanning electron micrograph of a fracture surface of a cured low-profile polyester resin filled with hollow glass micro-balloons (Fillite). (Average thickness of the glass wall of the filler particle in the bottom right-hand corner is 5 μm.)

is enhanced by the addition of the CTBN. Figure 18.2b gives results of fracture tests on cured samples of the same polyester resin, with and without the addition of Fillite particles, for two different grades of CTBN rubber. The CTBN 1 is Hycar CTBN X8 grade, while CTBN 2 is the more polar Hycar CTBN X31 grade. The results show the expected toughening effect of the reactive rubber on the polyester resin alone. The toughening actions of the

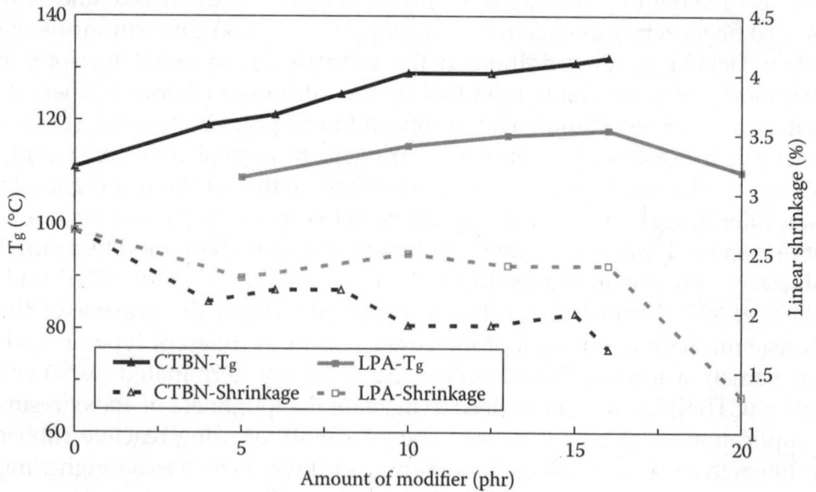

FIGURE 18.2A

Dependence of the linear shrinkage α and the glass transition temperature T_g of a polyester resin upon the type and amount of modifier (expressed in parts per hundred parts or resin, by weight).

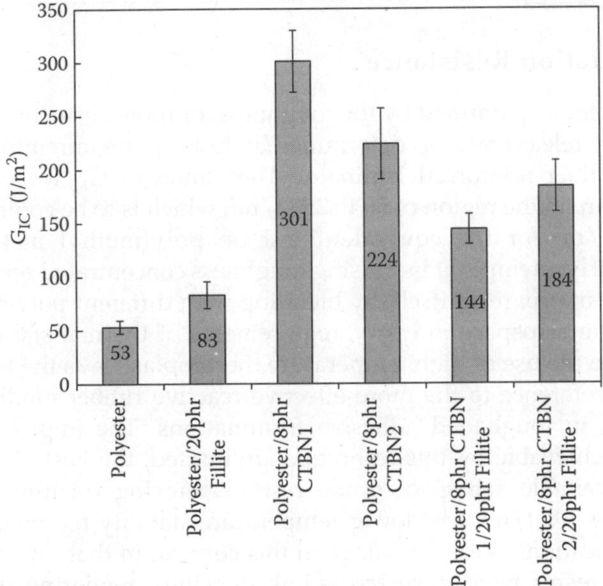

FIGURE 18.2B

Dependence of the toughness G_{IC} of a polyester resin with and without 20 phr of Fillite inorganic filler upon the type of CTBN modifier.

inorganic filler and the rubber are not synergistic, but it is clear that a suitable combination of toughness, density, and cost of the final cured blend may be reached.

Improving Delamination Resistance

In continuous fiber reinforced laminates, the planes between the individual prepreg plies are relatively weak, delaminating even under modest levels of through-the-thickness loading. This low resistance to delamination has represented an acceptance barrier for such materials in structurally critical applications, and delamination cracking is a universal concern in materials selection and structural design. For automotive applications, there is a direct read-across from the approaches to limiting of this delamination that have been explored by the aerospace composites industry over the last three decades. For the purposes of quantifying the delamination resistance achieved by different toughening approaches, it is convenient to split the problem into the crack initiation and crack propagation stages of fracture.

Crack Initiation Resistance

This parameter is quantified by the toughness, or more correctly, the critical strain energy release rate G_{IC} determined by tests on the unreinforced resin[5] and/or the fiber reinforced laminate.[6] The values of G_{IC} for unmodified thermosets are in the region of 50 to 250 J/m², which is to be compared with some 500 J/m² for the equivalent test on poly(methyl methacrylate) (PMMA). Early attempts at increasing toughness concentrated on the toughening of the matrix resin itself, by blending with different polymeric modifiers.[7,8] In the aerospace industry, requirements of thermal stability above 120°C dictate the use of high temperature thermoplastics as the toughening agents, in preference to the more effective reactive rubber modifiers commonly used in toughened adhesive formulations. The improvements in toughness achievable by this technique are limited, the best of such commercially available aerospace grade resins achieving fracture toughness values below 500 J/m². The lower temperature stability required for automotive applications is an advantage in this context, in that lower temperature curing resins have lower cross-link densities, rendering them more toughenable.[9]

Unfortunately, at G_{IC} above approximately 400 J/m², the transfer of any further resin toughness improvement to increase toughness in a laminate is very poor, as the toughening mechanisms in the resin tend to be inhibited by the proximity of the stiff fiber plies.[10,11] The use of selective placement of toughened resin inter-layers in composite lay-up was a natural follow on from resin toughening itself. Placement of a resin-rich layer capable of a significant plastic deformation in the path, if predictable, of a delamination crack, can result in a significant increase in the delamination initiation resistance of that sample, with a minimal loss in stiffness. The improvement is particularly notable if the sample is loaded in forward shear, in the so-called Mode II fracture resistance, as shown in Figure 18.3. The barriers to the use of this technology in real structures are the need for very accurate predictions of regions of probable crack initiation and crack growth, and an increase in the manufacturing complexity. A compromise solution has been the use of polymeric particles, sprinkled onto the outer layers of prepreg plies, acting as fixed spacers and defining the minimum separation of composite plies, and hence, of the (regularly spaced) resin-rich layers. Several such prepreg systems have been very successful in recent years, both in the aerospace and in the automotive industries. Figure 18.4 is a micrograph of one such system, clearly showing the polymeric spacer particles. High increases in delamination initiation resistance and compression-after-impact resistance can be achieved in composites made in this way, but at the expense of reduced specific stiffness. A more detailed treatment of selective resin interleaving is available in a recently published case study.[12]

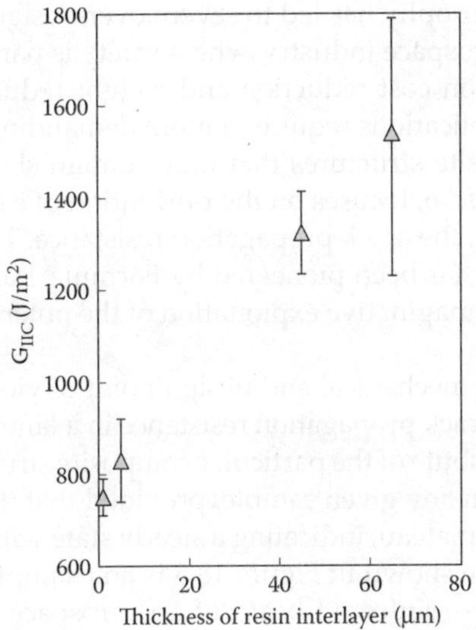

FIGURE 18.3
Increase in the Mode II delamination fracture resistance (*crack initiation*) in a composite beam, as the thickness t_m of a resin-rich central layer is increased.

Crack Propagation Resistance

For many years, the initiation of a crack in a composite structure has been considered as the final failure of that structure. Indeed, the only criterion obtainable from the ISO 15024 delamination fracture standard is that of crack

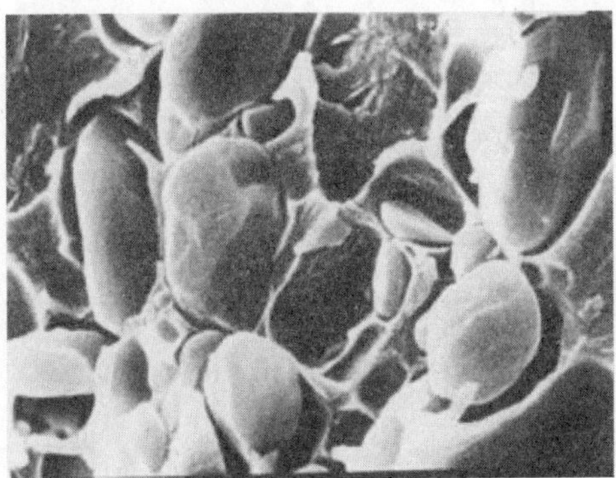

FIGURE 18.4
Polymeric spacer particles in a toughened epoxy resin, revealed by scanning electron microscopy of a cryo-fracture of a sample of the cured resin. (Particle size range 1 to 10 μm.)

initiation. This philosophy has led to severe over-design of composite structures in the civil aerospace industry, where safety is paramount. The current growing emphasis on cost reduction and weight reduction in many structural composite applications requires a more demanding appreciation of the behavior of composite structures that may contain short, but finite cracks. The attention, therefore, focuses on the prevention of catastrophic growth of any such cracks, i.e., the crack propagation resistance. The damage tolerance approach to design has been pioneered by Formula 1 applications that continue to lead in the imaginative exploitation of the potential of new materials and techniques.

From the fracture mechanical and design point of view, the basic problem lies in the fact that crack propagation resistance in a laminate is not a material property, but an attribute of the particular composite structure. The parameter may be quantified in any given sample, provided that the crack propagation resistance reaches a plateau, indicating a steady state with respect to the crack growth. The R-curve shown in Figure 18.5 is an example of data from a test of a unidirectionally reinforced beam of an aerospace grade carbon fiber/epoxy laminate. The crack is initiated from a thin polymeric starter film, and the resistance to crack propagation more than doubles as the crack extends along the beam. Such stabilization of crack growth is attributable to the crack bridging action of reinforcing fibers that span the layers immediately surrounding the delamination plane. Some fiber bridging may be obtained in

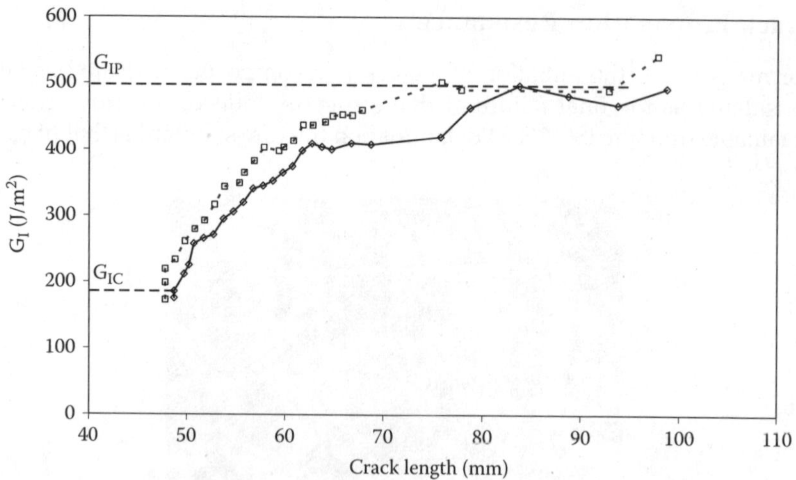

FIGURE 18.5

Delamination resistance against crack length (R-curve) obtained from a Mode I delamination test carried out on a unidirectional sample of IMS/924 carbon fiber/epoxy laminate. G_{IC} designates the critical strain energy release rate required to initiate a delamination crack from a thin starter film; G_{IP} indicates the limiting value of crack propagation resistance in the particular sample. While G_{IC} is regarded as a material parameter, the value of G_{IP} is known to depend strongly on the geometry and the mesostructure of the particular sample.

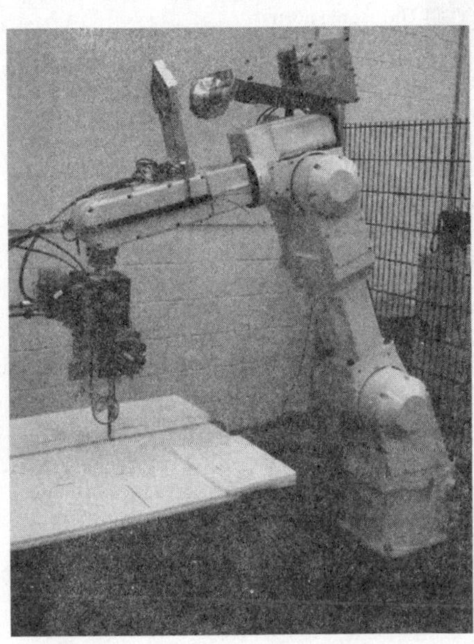 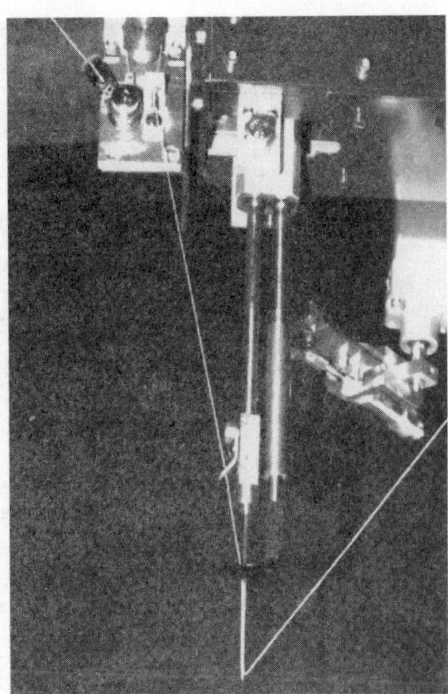

FIGURE 18.6
Tufting head (KSL[15]) installed on a Kawasaki robot arm (left) and detail of the tufting needle and thread arrangement (right).

any laminate from misaligned and nested reinforcing fibers and tows, but it is now generally agreed that the most effective way of delaying, or even preventing, delamination failures is by the use of deliberately introduced z-direction (through-the-thickness) reinforcement. Stitching with thin synthetic fibers was the first such technique to be explored. Very significant improvements in damage resistance of woven fabric composites were achieved, for a limited loss in the in-plane properties of the composites.[13,14] Tufting is the most recent version of this technology: a loop of tufting thread is introduced into the fibrous preform by a tufting needle; the needle extracted following the same path, is then moved to the next predetermined position, and the process is repeated.[15] The process relies on the friction between the thread and the fibrous preform to keep the tufts in position; once a successful match is achieved, the procedure is highly suited to automation. Figure 18.6 shows a tufting head installed on a robot arm in the author's laboratory, and a more detailed photograph of the needle/thread arrangement.

The two-sided or single-sided stitching technique is seen to be most applicable as a global reinforcement for flat panels, or a more local reinforcement for more complex fiber preforms. The composite structure is then achieved by liquid resin filling of the preform and out-of-autoclave cure. It is not really feasible to use any form of stitching with prepreg-based composites, as two-sided access to the lay-up is required, and the resin present

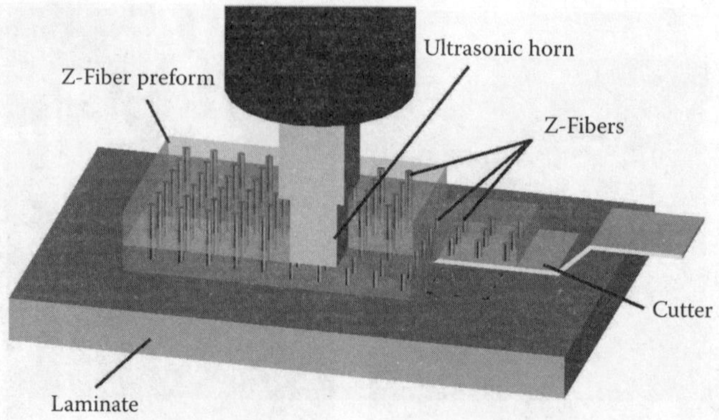

FIGURE 18.7
Schematic of the Z-pinning process: The Z-Fiber® preform is located on top of the uncured laminate, directly above the area to be reinforced. The Z-Fibers® are inserted into the laminates by the actions of the ultrasonic horn. During insertion the low-density foam holding the pins is crushed. The pins are shear cut at the surface of the laminate and the excess pin length and the remaining foam are discarded.

in the prepreg fouls the stitching needle. The most recently introduced through-the-thickness reinforcement technique of Z-Fiber® pinning (Z-pinning) overcomes this problem.[16–18] The stiff Z-Fibers®, hereafter referred to as Z-pins, are inserted orthogonally to the plane of the composite plies during the manufacturing process, before the resin matrix is cured, effectively pinning the individual layers together. The Z-pins can be made of metal, glass, or even ceramic, but their most convenient form for applications of the type considered here is as fully cured carbon fiber rods with an epoxy of bismaleimide matrix resin. The insertion into the **uncured** laminate lay-up is assisted by ultrasonic vibrations, as shown in Figure 18.7, and requires one-sided, on-mould access, resulting in much less damage to the fibers in the composite plies than stitching. Figure 18.8 shows how the Z-pins, pulling out from the composite under predominantly Mode I loading, stabilize the failure in a composite structural element, namely a T-joint subjected to a stiffener pull-off test.

This new technology is envisaged to be used as selective reinforcement, and has already been utilized to provide urgently needed solutions to increased load-bearing capacity regulations for several Formula 1 racing teams,[19] as well as answering significant structural integrity challenges in the case of military jets. The increase in the crack propagation resistance of laminates, offered by Z-pinning, is an order of magnitude higher than that obtainable by use of toughened resins.[20] Preliminary impact damage studies suggest that the Z-pins are not effective in preventing the early initiation of delamination cracking, but are highly effective in reducing and stopping subsequent crack growth. Thus, the most damage tolerant composite structures of the future might be obtained by a combination of

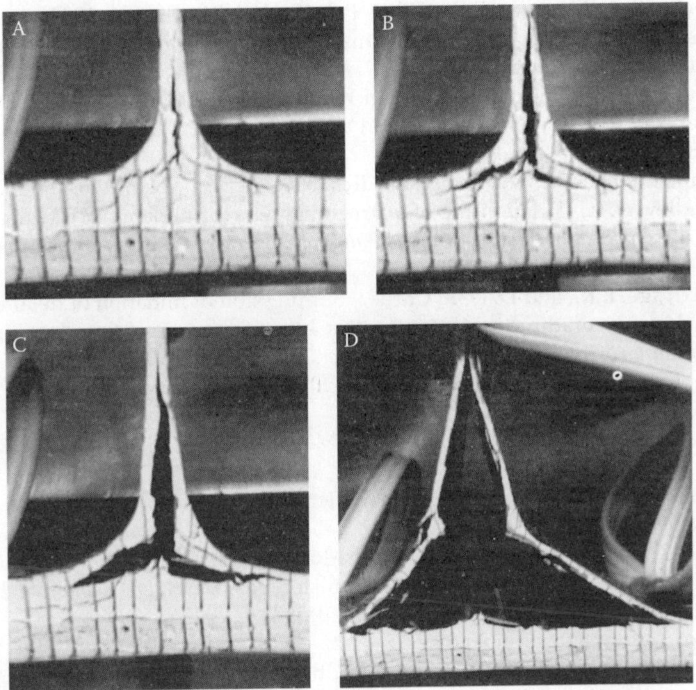

FIGURE 18.8
Progressive failure in a T-pull off test of a composite T-joint, reinforced by localized Z-pinning in the stiffener foot region. The vertical displacements are 0.05, 0.13, 0.62, and 3.39 mm, going from (a) to (d). It should be noted that the equivalent, but unpinned, joint failed catastrophically at vertical cross-head displacement of 0.13 mm (b). In (d) the carbon fiber Z-pins pulled out of the stiffener foot are visible. (The vertical lines marked on the side of the specimen are for purposes of crack length determination only.)

selectively placed crack initiation resistant resin layers with Z-pinned crack stopping regions.

References

1. Pattison, V. A., R. R. Hindersinn, and W. T. Schwartz, *J. Appl. Plym. Sci.*, No. 18, 2763, 1974.
2. Bucknall, C. B., I. K. Partridge, and M. J. Phillips, *Polymer*, No. 32, 636–640, 1991.
3. Li, W., and L. J. Lee, *Polymer*, No. 39, 5677–5687, 1998.
4. Kinloch, A. J., *Structural Adhesives: Developments in Resins and Primers*, A. J. Kinloch, ed., Elsevier Applied Science Publishers, London, 127–162, 1986.
5. ISO 13586 *Determination of Fracture Toughness (Gc & Kc) for Plastics.*
6. ISO 15024 *Determination of Mode I (Critical Strain Energy Release Rate or Fracture Toughness, GIC) of Unidirectional Fibre-Reinforced Polymer Laminates Using the Double Cantilever Beam (DCB) Specimen.*

7. Bucknall, C.B., and I. K. Partridge, *Polymer*, Vol. 24, 639–644, 1983.

8. Pascault, J.-P., and R. J. J. Williams, "Formulation and Characterization of Thermoset-Thermoplastic Blends," *Polymer Blends*, Vol. 1, Chapter 13, D. R. Paul and C. B. Bucknall, eds., John Wiley & Sons, 2000.

9. Yee, A.F., and R. A. Pearson, *NASA Contractor Report*, p. 3718, 1983, and 3852, 1984.

10. Hunston, D. L., *ASTM Composites Technology Review*, Vol. 6, 176, 1984.

11. Bradley, W. L., "Relationship of matrix toughness to interlaminar fracture toughness," *Application of Fracture Mechanics to Composite Materials*, Chapter 5, K. Friedrich, ed., Elsevier, 1989.

12. Partridge, I. K., and D. D. R. Cartié, "Suppression of initiation of delamination cracking in unidirectional composites by self-same resin interleaving," *The application of fracture mechanics to polymers, adhesives, and composites*, D. R. Moore, ed., ESIS Publication 33, ISBN 0 08 044205 6, Elsevier Applied Science Publishers, London, 2003.

13. Dransfield, K. A., L. K. Jain, and Y.-W. Mai, *Comps. Sci. & Tech.*, Vol. 58, 815–827 and 829–837, 1998.

14. Mouritz, A. P., K. H. Leong, and I. Herszberg, *Composites: Part A*, Vol. 28A, 979–991, 1999.

15. KSL GmbH, website: http://www.ksl-lorsch.de.

16. Tong, L., A. P. Mouritz, and M. Bannister, eds., *3D Fibre Reinforced Polymer Composites*, Elsevier Science, ISBN 008 043938-1, 2002.

17. http//:www.aztex-z-fiber.com.

18. Partridge, I. K., D. D. R. Cartié, and T. Bonnington, "Manufacture and performance of Z-pinned composites," *Advanced Polymeric Materials: Structure-Property Relationships*, Chapter 5, S. Advani, and G. Shonaike, eds., CRC Press, Boca Raton, FL April 2003.

19. "Safety pins," *Journal of Racecar Engineering*, 56–62, December 2002.

20. Cartié, D. D. R., and I. K. Partridge, "Delamination behaviour of Z-pinned laminates," *Proceedings of 2nd ESIS TC4 conference*, (Les Diablerets, Switzerland, 13–15 September 1999), J. G. Williams and A. Pavan, eds., Elsevier, ESIS Publication 27, ISBN 008 043710-9, 2000.

19

High-Performance Autosport Surface Treatments and Composites

Roger Davidson, Ed Allnutt, and Will Battrick

CONTENTS

Lightweight Materials Options

In the automotive sector, the wider use of light alloy materials such as aluminum and magnesium, and structural composites offers an undeniably attractive means of gaining greater performance with increased fuel economy and reduced emissions. Within the realm of autosport, the desire to minimize weight is even more pressing—those involved in engineering the cars and bikes that compete every week strive to utilize high-specific strength and stiffness materials in as many areas as possible. This chapter will consider the use of environmentally friendly enhanced surface treatments on light alloys, as well as some specific applications for highly stressed lightweight composite components.

Surface Engineering of Light Alloys

Until recently, the widespread use of light alloys in autosport has been hindered by the surface performance of these materials. This is in contrast to their specific bulk properties, which have been improved to the stage where they surpass their ferrous competitors. With its very low specific gravity of ~1.8, magnesium alloys are particularly attractive.

Applications for magnesium alloys have been limited by their reputation for susceptibility to corrosion. In fact, degradation of magnesium is closely related to its purity with respect to contaminants such as copper, nickel, and iron. Close purity control can give magnesium components salt spray corrosion behavior better than many aluminum alloys.

In recent times, the metallurgy of the magnesium system has been subject to a great deal of improvement, and many new and potentially beneficial surface modification technologies have begun to reach commercialization. These have brought enhancements to corrosion behavior, tribology, and aesthetic appearance, while simultaneously overcoming the environmental restrictions of the early systems.

The most recent developments, centered on the phenomenon of plasma electrolytic oxidation (PEO), offer exceptional performance enhancements combined with an opportunity to develop exotic surfaces through second-phase infiltration. Plasma electrolysis is a relatively new electrochemical treatment process/discipline. The most important derivative is plasma electrolytic deposition (PED), which includes plasma electrolytic otidation and plasma electrolytic saturation (PES). In plasma electrolytic deposition, spark or arc plasma micro-discharges in an aqueous solution are utilized to ionize gaseous media from the solution, such that complex compounds are synthesized on the metal surface through the plasma chemical reactions.

Keronite™ Hard Ceramic Surfacing for Light Alloys

Keronite ceramic surfacing for aluminum, magnesium, and titanium alloys is a new environmentally friendly electrolytic conversion coating that effectively substitutes traditionally more hazardous processes, such as anodizing and chromating.

Keronite is a corrosion resistant, hard (400–1900 HV) oxide of the substrate material. Keronite thickness ranges from 5 to 150 microns. The layer is predictable, uniform, and well bonded with the metal substrate. High hardness and toughness of the Keronite surface enables aluminum and magnesium alloys to be used in new applications. The process was developed and commercialized by Keronite Limited, United Kingdom. Crompton Technology

Group (CTG) has a manufacturing licence and has been developing the process and assessing treated components in a number of automotive and other industrial applications.

Keronite Process

Ceramic layers produced on aluminum, magnesium, and titanium by a novel plasma electrolytic oxidation technology offer an attractive alternative to chromating, hard anodized, and conversion coatings in wear-resistant and corrosion-resistant applications. Composite and multilayer coatings incorporating lubricious or additional hard phases can also be produced. Like anodizing, the process uses an electric power supply and a bath, but it is significantly different from anodizing because it produces harder and thicker layers while using ecologically compatible alkali electrolytes, and a specially modulated AC power supply.

Preparation for the Keronite coating process requires only de-greasing of the part. No special pre-treatment of the surface is required because the Keronite process makes use of electric micro-discharges in the electrolyte that cleans the surface. Coating usually takes 10–60 minutes depending on the required thickness. The coating grows at the interface, and depending on the relative density of the base alloy to the oxide layer, the specimen dimensions are increased. After treatment, the part is rinsed in warm water for several minutes. The low concentrated alkali electrolyte does not contain any toxic or aggressive elements and is no more hazardous than water in a washing machine. Figure 19.1 shows the Keronite coating apparatus.

FIGURE 19.1
Keronite coating apparatus.

Typical features of the Keronite coatings are described below:

Extremely hard and wear-resistant: hardness ranges from 350 to 600 HV (Rc = 36 – 55) on magnesium and 600–1900 HV on aluminum. In two body abrasive wear situations, Keronite reduced the wear rate of uncoated magnesium alloys by a factor of 20 an aluminum alloys by ~100.

Low-friction: once polished, the coating has a friction coefficient of less than 0.15 (against steel). Keronite can eliminate the high friction and galling normally associated with magnesium.

Corrosion resistance: outdoor tests show months without corrosion.

Coating thickness: thicknesses from 10 μm to 80 μm provide a uniform surface layer.

High dielectric strength: the coatings withstand in excess of 1000 V DC, and are therefore well-suited as surfaces for electrical components.

Heat resistance: the coatings can typically withstand short exposures of up to 1000°C.

In wear-resistant applications, Keronite coatings can be used at thicknesses up to ~150 μm on aluminum and titanium alloys, and ~80 μm on magnesium alloys, and layers 200–600 μm in thickness can be produced where corrosion resistance or electrical insulation is the main requirement. The process allows a uniform layer to be formed on complex shapes without cracking, and little surface preparation is required. Worn or damaged parts can be recoated without stripping the damaged coating. Coating of deeper holes or bores may require special jigging and electrodes. Best results are obtained on aluminum alloys containing <10%Si, but—unlike with hard anodizing—other alloying elements, such as copper, are beneficial to coating hardness and wear resistance. Best results on titanium and magnesium are obtained on alloys, rather than the pure metal. Some typical coating thicknesses and hardnesses are given in Table 19.1, although microhardness levels up to 1900 HV have been achieved on 2xxx series aluminum alloys.

TABLE 19.1

Properties of Typical Keronite Coatings on Various Alloys

Alloy	Thickness (μm)	Microhardness, HV (100 g)
AA 6082	25–250	1250
A 7075	25–200	1350
A 2024	25–150	1550
Ti 6-4-1	10–100	500
Mg AZ91	5–50	350

Coating Structure

The Keronite layer is the complex oxide ceramic consisting of hard crystalline phases dislocated in the matrix of softer phases of oxide. Such structural composition provides Keronite with a combination of high hardness, wear resistance, and shock resistance. These attributes explain the higher flexibility of Keronite, as well as nearly three times less reduction of basic material fatigue strength compared with anodizing.

Keronite coating porosity gradually decreases with depth. The porosity is a maximum at the surface of 15%–30% and is characterized by the pores of typical microns. The ceramic layer close to metal substrate is much less porous at 2%–10%. The porosity can be controlled by the AC voltages applied during the treatment process. Unlike anodized surfaces, Keronite porosity is randomly distributed throughout. The graded structure of porosity can enable subsequent impregnation of the Keronite surface with a variety of materials, such as paints, organic materials, metals, carbides, and even diamond-like carbon.

The Keronite process involves conversion and enrichment of the surface region, rather than deposition of a new layer. Because the coating is formed through a reaction that involves the substrate, adhesion is excellent. Surface growth is approximately 30% of the coating thickness. As deposited, the coating consists of a thin transitional layer between substrate and coating, a compact, layer with low porosity, and an outer layer characterized by higher (>15%) porosity. For some applications, the outer layer should be ground or polished away, but the porosity can provide keying for further layers applied to optimize performance. Secondary treatments include coating with PTFE to reduce friction, polymer sealants to increase corrosion resistance, and impregnation with nickel, chromium, copper, or carbides.

Ball on disc tests have shown that the adhesive wear resistance of Keronite on aluminum alloys, particularly 2024, is superior to that of hard chromium layers, and an order of magnitude better than hard anodizing or electroless nickel coatings. Coated 6082 aluminum alloy has survived over 2000 h in salt spray tests, but high copper containing alloys perform less well (200 h), although their performance can be improved by polymer sealing. The coatings exhibit good thermal resistance, and coatings can withstand short exposures up to 2000°C, which makes them ideal for parts, such as rocket venturi, and prolonged exposure to high temperatures has no adverse effect on the coating performance, other than some crazing of the surface. Keronite layers also provide suitable substrates for mounting electrical components. Breakdown voltages are typically 10 V μm^{-1}, and can be increased further by impregnation.

Keronite Composites

The Keronite surface can be used as a matrix for composite coatings. Two coating systems can be used in conjunction, in order to create superior properties for the combined coating. External coating partly penetrates the porous surface of Keronite and partly remains on the surface. The penetrated region comprises a composite material that is characterized by high hardness provided by the Keronite ceramic, and by the properties of impregnated materials, such as high-wear, corrosion resistance, lubricity, and electrical resistivity. Molybdenum carbide, diamond-like carbon, electroless nickel, polymers, and lacquers can all be applied. Keronite ceramic coatings in combination with even harder materials, such as metal carbides dislocated on and in its external layer, enables the creation of super-hard coatings well-bonded with the substrate.

Applications

The properties provided by a Keronite surface on magnesium alloy enables their use in all traditional magnesium alloy application areas. Figure 19.2a shows a cross-section through a typical Keronite coating on a magnesium alloy, whereas Figure 19.2b shows a coating on an aluminum alloy. Figure 19.2c shows a typical Keronite surface. The High hardness and toughness of the Keronite surface enables magnesium alloys to be used in many new functioning applications, as illustrated in the following examples:

- Improves wear of "strong" magnesium alloys such as WE54, and widens their range application in engines, for example, pistons, loaded elements of pumps, fuel and pneumatic drives, and sliding bearings.

- Improves wear resistance of lightweight castings, providing a bearing surface for high performance engine oil pumps, hydraulic systems, etc.

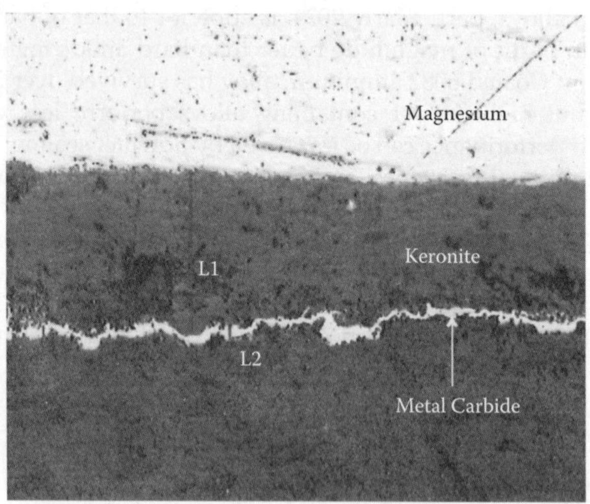

FIGURE 19.2A

A cross-section of a Keronite coating on a magnesium alloy.

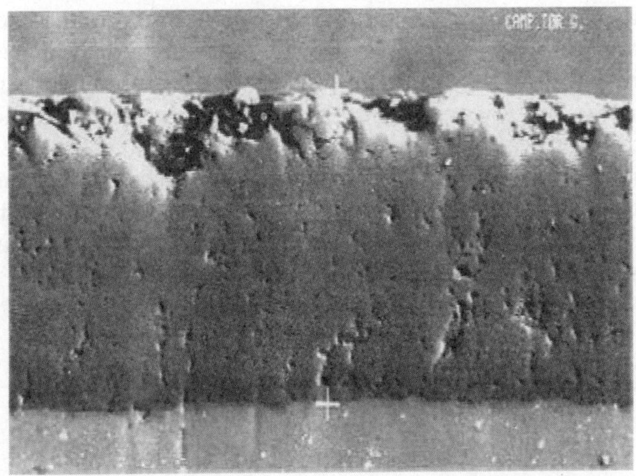

FIGURE 19.2B
A cross-section of a Keronite coating on an aluminum alloy.

- Stiffens ultra-thin alloy sections and fabrications, improves corrosion protection, for example, covers, housings, pressure die-casting, and manifolds.

- In electronic applications, Keronite provides a hard, scratch-resistant undercoat for paint or lacquer, and can be used for mobile phones, lap-tops, camcorders, and camera applications.

- Some coating technologies, such as plasma spraying, may result in degradation to the underlying component. The application of Keronite as a thermal barrier and abrasion resistant treatment for

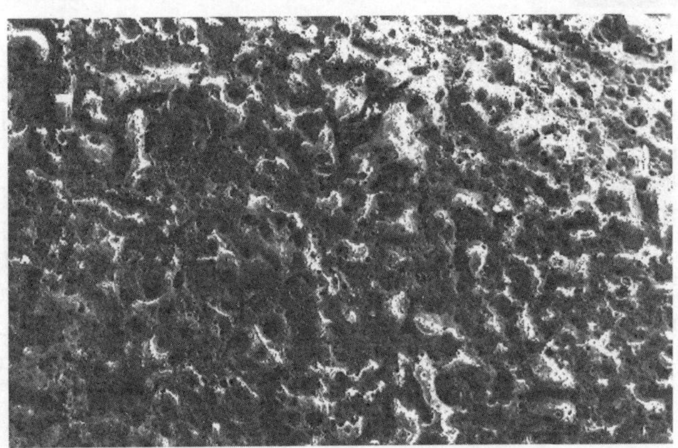

FIGURE 19.2C
A typical surface of a Keronite coating.

racing pistons may provide a performance advantage because of greater adherence, and improved thermal performance with a corresponding knock-on reduction in the required cooling system capacity. There can also be a reduction in carbon build-up in piston applications; a reduction in thermal load on the piston, which in turn, allows a reduction in piston size through improved materials properties at lower temperatures. Non-autosport applications in pistons are likely in the treatment of aluminum alloy diesel pistons where piston design can be enhanced by the use of Keronite coatings.

Polymer Composites

Polymer-based composites with carbon and glass fibers have been used in automotive applications for lightweight body panels. However, materials and manufacturing costs and, more recently, disposal and recycling, have restricted their greater use. Nonetheless, one area that has been receiving more interest recently has been the use of natural fiber reinforced composites designed for biodegradability at end of life. These materials will be even more attractive once they have been combined with biodegradable, but environmentally stable, natural polymer-based matrices. At the end of their life, these systems degrade when composted. These types of systems will not be considered further in this chapter, but instead the focus will be turned on much more mature highly stressed composites used in motorsport.

Highly Stressed Composites

Propellor shafts (or prop-shafts) take power from the engine to the gearbox and are used in motorsport applications worldwide. CTG's composite TORQline™ range of prop-shafts have helped teams to achieve success at Le Mans, in FIA GT, ALMS, and other series. The approach to the production of prop-shafts begins with the design of the optimized composite tube element. Each tube is a bespoke design for the application, and is manufactured by filament winding, or in some cases, tube rolling. These hybrid tubes are designed in conjunction with the end fittings, and bond to ensure that they work together to give the strongest and most reliable solution. Options for couplings include composite discs, tripodes, and Hooke's joints.

The basic attraction of composite materials for prop-shafts (or any long driveshaft application) is that they make it possible to increase the critical speed or bending resonant frequency f_c of the shaft according to:

$$f_c = k \frac{d}{L^2} \sqrt{\frac{E}{\rho}} \tag{19.1}$$

If the length L that the shaft needs to span is fixed, and the diameter d is restricted, then the only way of improving the critical speed is to increase the modulus E divided by density ρ (the specific modulus E/ρ). The most effective way of improving the specific modulus is to use carbon fiber composite, and this enables dramatically longer and lower weight single-piece driveshafts to be produced. For very long shafts, low angle, high stiffness pitch fibers can be incorporated to increase the whirling speed. Higher angle fibers react the torque and increase torsional buckling resistance. Typical properties are given in Table 19.2.

Composite torque transmission shafts offer significant weight savings over more traditional materials without any decrease in mechanical properties. The reduction in mass has a direct impact on the force required to accelerate/decelerate the vehicle.

The advantages of using composite prop-shafts may include:

- **Light weight.** Typical weight savings of up to 50%. This reduces the overall kerb weight of the vehicle and, therefore, the force required to accelerate and decelerate.

- **Low mass moment of inertia.** This is a measure of the rotational inertia of a part, and its resistance to change in rotational velocity. Due to the low density and high mechanical properties of carbon fiber reinforced polymer prop-shafts, the moment of inertia is significantly less than steel prop-shafts, improving the overall vehicle performance.

- **Whirl elimination.** Due to their low cross-sectional area and mass, and high longitudinal stiffness, composite prop-shafts have higher whirling frequencies than other materials. This allows composite prop-shafts to span large lengths at high speeds. CTG has experience of supplying single-part prop-shafts to replace directly two-part shafts, offering significant weight savings and eliminating the use of a center bearing.

TABLE 19.2

Typical Prop-Shaft Properties

Property of the Prop-Shaft	Value
Weight	2.0 kg
Length	1.5 m
Diameter range	58.5 to 75 mm
Temperature range	<120°C peak, <100°C operating
Failure torque	3500 Nm
Fatigue life	>10^6 Cycles @ 85% failure torque

- **Crash-worthiness.** Due to the mode of failure in large impact collisions, composite prop-shafts are more crash-worthy, and safer than other materials. Steel prop-shafts have a tendency to penetrate floor panels. In comparison, composite shafts tend to delaminate and crush.

- **Torsional stiffness.** Through the use of different materials or different wind patterns, the torsional stiffness of the shaft can be tailored and balanced against the bending stiffness. Composite prop-shafts can be used to add compliance into the driveline. Composite shafts can also be manufactured to stiffness match other materials, such as composite anti-roll bars.

- **Robustness.** The basic carbon fiber reinforced polymer structure can be protected against the effect of stone impact by use of toughened overwound layers of polymer-based fibers. Temperatures of the bonded ends should, however, be restricted to <90°C.

Figures 19.3a and 19.3b show the filament winding of a carbon fiber reinforced polymer prop-shaft and the testing of the finished component, respectively.

Other similar applications in motorsport applications include driveshafts, anti-roll bars, strut braces, steering columns, motor shafts, and torque tubes.

Although most applications to date have been in motorsport/rally applications, large potential benefits are anticipated in heavy transport applications for lorries and commercial vehicles. Other applications of composite shafts include helicopter and marine shafts, as well as industrial shafts for paper making and printing industries.

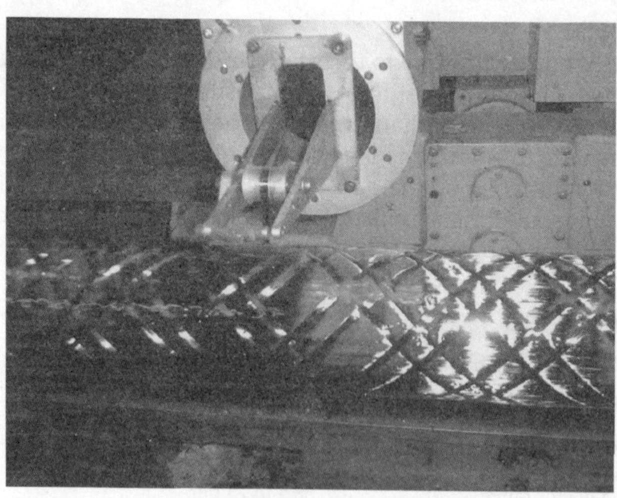

FIGURE 19.3A
Filament winding of a CFRP prop-shaft.

FIGURE 19.3B
Testing of assembled CFRP prop-shaft.

Driveshafts transfer the torque to the wheels and a system has been developed for the production of carbon fiber reinforced driveshafts that can have either Tripode end fittings integrated into the structure, or splined end fittings. A shaft of typically 35 mm can be used to transmit a torque of > 3000 Nm, and offers a weight saving of 40%–50%. Each shaft is designed to meet each application in order to give maximum benefit to each customer.

Light-Weight Pressure Vessels

Another potentially very important application for highly stressed composites is in lightweight pressure vessels and accumulators for the storage of gases and fuels. CTG's experience in pressure vessel design and manufacture ranges from small pressure vessels of volume 200–700 cc for use as nitrogen vessels for F1 valve train applications to large vessels of 400 mm diameter and 1500 mm length for sub-sea applications to depths of over 4000 m, as well as large spheres for storage of oxygen pressurant. Most vessels require a metallic or polymer liner.

A new dynamic liner technology has been developed that gives these types of vessels exceptional fatigue resistance, allowing both low shape profile reservoirs and accumulators for active systems to be produced. These vessels offer a versatile design, high strength, low weight, and excellent

FIGURE 19.4
Filament winding of a CFRP accumulator vessel on a thin-walled aluminum liner/mandrel.

fatigue resistance. The flexibility of the design also allows bladders and diaphragms to be incorporated into the liners. Figure 19.4 shows the filament winding manufacture of a CFRP accumulator on thin-walled aluminum mandrel.

These high-performance vessels can be produced to virtually any size, with geometries made up of cylinders, spheres, domes, and cones. Significant weight savings can be made as shown in Table 19.3.

This type of technology will become much more important in transport applications for the storage of high-pressure hydrogen as fuel cell technology becomes more refined.

TABLE 19.3

Typical Properties of Composite Pressure Vessels

Property	Value
Weight saving	*60%, **40%
Operating pressure	220–400 Bar
Burst pressure	650–1500 Bar

*versus Al vessel

**versus Ti vessel

Conclusions

Advanced materials and surface technologies have always played a critical role in the optimization of autosport vehicles. These types of technologies will become increasingly important in the commodity automotive vehicle sector, as environmental concern and emissions reduction gain even greater international priority. From a small company perspective, this chapter has attempted to focus on some noteworthy highlights, and should help point to directions for future collaborative efforts.

Index